Commercial & Regional Transport Aircraft Pilot Reports

Other books by *Aviation Week & Space Technology* Magazine

Military Aircraft Pilot Reports
Business & General Aviation Aircraft Pilot Reports
Helicopter Pilot Reports

Commercial & Regional Transport Aircraft Pilot Reports

Aviation Week & Space Technology Magazine

McGraw-Hill

New York San Francisco Washington, D.C. Auckland Bogotá
Caracas Lisbon London Madrid Mexico City Milan
Montreal New Delhi San Juan Singapore
Sydney Tokyo Toronto

The MD Explorer on the cover is the first new twin-turbine commercial helicopter for the twenty-first century, according to its manufacturer, McDonnell Douglas in Mesa, Arizona. The company claims that the Explorer is the safest, quietest helicopter in its class, featuring the NOTAR (no tailrotor) antitorque system, on-board health monitoring of aircraft systems, and seating for eight. McDonnell Douglas photograph by Robert W. Ferguson.

McGraw-Hill

A Division of The ***McGraw-Hill*** *Companies*

Published by The McGraw-Hill Companies, Inc.

pbk 1 2 3 4 5 6 7 8 9 DOC/DOC 9 0 0 9 8 7 6 5

Library of Congress Cataloging-in-Publication Data

Commercial & regional transport aircraft pilot reports / by Aviation week & space technology magazine.
p. cm.
ISBN 0-07-003167-3 (pbk.)
1. Jet transports—Evaluation. 2. Jet transports—Piloting.
I. Aviation week & space technology.
TL685.7.C677 1995
629.132'5216—dc20 95-36858
CIP

Acquisitions editor: Shelley IC. Chevalier
Editorial team: Robert E. Ostrander, Executive Editor
Sally Anne Glover, Book Editor
Production team: Katherine G. Brown, Director
Susan E. Hansford, Coding
Janice Ridenour, Computer Artist
Toya Warner, Computer Artist
Rose McFarland, Desktop Operator
Jodi L. Tyler, Indexer
Design team: Jaclyn J. Boone, Designer
Katherine Lukaszewicz, Associate Designer

GEN1
0031673

Contents

Introduction

The cockpits of commercial transport aircraft were transformed in the early 1980s by the advent of digital avionics, glass cockpit displays and automated flight decks for two-man crews as the Boeing 757 and 767 entered service. Then in the late 1980s, even higher levels of automation were achieved with fly-by-wire flight controls in the Airbus Industrie A320.

The pilot reports in this section cover this critical transition period in commercial aviation. Aviation Week & Space Technology pilot/editors wrote about their hands-on encounters with the new cockpits that marked the end of one era and the beginning of the next. The stories are written with line pilots in mind, pilots who were trained in an analog era, but who will have to adapt to a digital way of life on the flight decks described here. The round dial, electromechanical instruments that dominated the post World War II era are on their way out, while Electronic Flight Instrument Systems (EFIS) and Engine Indication and Crew Alerting Systems (EICAS) are on their way in.

Then there is the flight management system computer and its related input/output device, the control display unit, which have become a critical part of the flight deck. There will be no turning back from the ubiquitous computer, but some pilots wonder if typing speed can really be more important than stick and rudder skills. Now in the mid 1990s, some airlines are again training pilots in recovery from unusual attitudes, so it appears that experience levels have come full circle. Flying skills still matter, even in a digital airplane with automated systems that can handle things from takeoff to touchdown. And clearly it is critical for the pilot to stay up to speed with the computer.

The authors of these pilot reports come from diverse aviation backgrounds in military and civil aircraft, which provides different perspectives on the latest technology in transport aircraft cockpits. The editors were invited to fly for 1-3 hours with test pilots, or other demonstration pilots highly qualified in the airplane. The intent of the reports is to cover normal aircraft operation and handling qualities and to include both low speed and high speed flight situations, such as approaches to stalls and in some cases actual stalls. But the focus of Aviation Week & Space Technology pilot reports has always been on line flying rather than more specialized testing that is often part of an aircraft certification program.

David North, the managing editor, is a carrier-qualified pilot who flew the A-4 Skyhawk during the Vietnam War. He also has line flying experience as a former Boeing 707 pilot/flight engineer with Pan American World Airways. North writes about the MD-90, the Airbus A320 and A340 and a variety of regional airline turboprops, including the de Havilland Dash 7 and 8, the ATR-42, Jetstream 41 and Saab 340. He has also written many pilot reports on U.S., Russian and European military aircraft which appear in another McGraw-Hill book.

Bruce Nordwall, Aviation Week's senior avionics editor, flew both the Grumman S-2 and Lockheed S-3 anti-submarine warfare aircraft in the U.S. Navy and logged 350 carrier landings and 5,000 hrs. He also served as the squadron commander of an S-3 unit before retiring as a captain with 26 years of service. Here he writes about the Fokker 50 and Fokker 100.

Edward H. Phillips, Aviation Week's transport editor, is a flight instructor and a commercial pilot with multi-engine rating and 2,000 hrs. Phillips worked for 14 years in the business aviation industry before joining the magazine. Here he contributes pilot reports on the Beech 1900D and the Cessna Grand Caravan.

Robert B. Ropelewski, a former member of the Aviation Week staff and before that a Marine Corps helicopter pilot who flew combat missions in Vietnam, writes about the Airbus A310, Boeing 757, 767, 737-300 and the Casa CN-235 turboprop.

My background as a former Air Force Reserve C-5A copilot has helped me in flying large transport aircraft, although the C-5A has few computers while the Airbus 330 and 321 and Boeing 777 are highly computerized, fly-by-wire transports. Since many pilots come to this new class of transports from a round dial background, it is easy for me to relate to the task they have in transitioning to a fly-by-wire aircraft. I also review the McDonnell Douglas MD-11, Canadair Regional Jet and Fokker 70.

Fly-by-wire controls provide many possibilities for the design engineer to assist the pilot. The demonstration of the Airbus alpha floor protection in the Airbus A330 was one of the most interesting experiences I have had on the three fly-by-wire aircraft I have flown. Pierre Baud, at that time the Airbus chief test pilot, was in command on the A330 demonstration flight in 1993 (see page 22). This is a maneuver I never even had the chance to see in the simulator in my days in the C-5A.

It is interesting to note that many of the airplanes reviewed here will be flying well into the 21st century and the latest transports will serve as a good model for the shape of things to come.

David Hughes
Northeast United States Bureau Chief

Chapter 1

Large commercial

Fly-by-wire 777 keeps traditional cockpit

David Hughes/Boeing Field, Seattle
May 1, 1995

The 777 cockpit is a cross between the 747-400's and 767's, but the similarities are only skin deep. The new flight control system depends on wires rather than cables. Boeing has combined a fly-by-wire system with traditional cockpit controls in the 777 to create a large, twin-engine transport that represents an evolution rather than a revolution in its product line.

Pilots of cable-controlled, glass-cockpit aircraft will feel right at home in the 777 cockpit. Its front panel is very similar to the one in a Boeing 747-400. The overhead panel and pedestal copy the 767 layout. There is a yoke that moves when the autopilot is engaged, and the throttles move when the autothrottle system adjusts power. This tactile feedback is an additional cue to the pilot as to what the autoflight systems are doing. (Airbus has taken a different approach on fly-by-wire cockpits by employing side-stick controllers and throttles that remain fixed when the autothrottle adjusts power.) Boeing also

The Pratt & Whitney turbofans on this demonstration flight were rated at 77,200 lb. of thrust each.

has added speed stability to the fly-by-wire logic so that forces build up on the control column when the aircraft changes speed. The pilot then neutralizes the forces with trim.

The internal structure of the 777 avionics system represents a dramatic departure from previous Boeing designs. Fly-by-wire computers warn pilots when they are flying too fast or slow, or when they are overbanking. Boeing does not limit the pilot's ability to override the fly-by-wire computer and command a high-g pull-up or to roll more than 60 deg. to recover from an unusual attitude. The Honeywell Airplane Information Management System integrates all avionics functions in two cabinets.

Boeing also has designed the 777 to perform extended-range twin-engine operation (ETOPS) with redundancy built into the electrical system. Simplified controls and displays will make it easy for pilots to divert to an alternate airport if necessary. The aircraft also handles well on one engine with no unusual control problems, even though up to 77,200 lb. of thrust is provided on one side of the aircraft. The 777 is quite agile for a heavy transport, providing rapid roll rates of up to 20 deg./sec. for evasive maneuvering.

This *Aviation Week & Space Technology* pilot had the opportunity to fly a 777 at Boeing Field last month. The flight test aircraft, sequence number WA001, made the first flight on June 12, 1994 (AW&ST June 20, 1994, p. 20). The aircraft has flight test monitoring equipment in the cabin. For our flight, the Pratt & Whitney turbofans provided 77,200 lb. of thrust.

John E. Cashman, chief pilot on the 777 program, occupied the right seat while I took the left seat. After we entered the cockpit, Cashman started the APU and switched power on the 777 electrical system, which made a smooth "no break" transition from ground cart power. The overhead scan of system status is very straightforward.

The aircraft's gross weight was 445,000 lb., including 135,500 lb. of fuel. Cashman set up our takeoff speeds in the flight management system (FMS) control display unit. V_1 would be 125 kt., rotate (V_R) would be 129 kt. and takeoff safety speed (V_2) would be 137 kt. The speeds are all marked on the vertical tape on the primary flight display. We would climb out initially at about 152 kt. (V_2 + 15) as commanded by the flight director when VNAV engages at 400 ft. above the ground.

The 777 has two 8-×-8-in. flatpanel, liquid crystal displays (LCDs) in front of each pilot—one for primary flight information and one for navigation. There is also one multifunction display (MFD) in the center of the panel and a second one below it in front of the throttle quadrant on the center pedestal. The large-format displays from Honeywell make even minute details easy to read on the navigation map display. The three small Rockwell Collins LCDs for standby attitude, speed and altitude are remarkably clear and easy to read.

When 777s enter service, they will also have data-link capability that will allow pilots to receive weather and company information for display on the MFD or to be printed out on a printer at the rear of the center pedestal. The data-link system also is designed to accommodate future ATC transmissions on speed, heading and altitude assignments.

A flight control synoptic page is new for Boeing aircraft, and it shows the movement of all of the primary flight controls—ailerons, flaperons, spoilers, elevators and the rudder. When control is lost over some surfaces due to a hydraulic system failure, that surface is simply crossed out on the display, making it easy for the pilot to see what he has left.

Cashman set up the navigation displays for a departure on Runway 31L at Boeing Field and a flight to the east to Grant County Airport at Moses Lake, Wash., for touch-and-goes. We planned to use full thrust on takeoff and climb with flaps at 20 with a center of gravity of 28% mean aerodynamic chord.

We used a paper checklist on our flight, since WAOO1 is not equipped with an electronic checklist. Earlier, I had the chance to see the electronic checklist work in both normal and emergency situations in the simulator. There are closed loop features in the checklist that show an item in green when it has already been accomplished.

For example, the system automatically checks flap handle position, flap position and the flight management computer-designated flap setting to see if they agree prior to takeoff. Some checklists involving abnormal situations have deferred items that are automatically appended to the appropriate checklist and presented to the pilot later. This is a nice electronic reminder.

We set the 777's auto-brake system on the RTO position, which means that any reduction of thrust to idle above 80 kt. on takeoff roll would result in a maximum brake application with the full 3,000 psi. available.

It is possible to start both of the Pratt & Whitney engines simultaneously, and Cashman placed both start/ignition switches and both fuel control switches to run. Airlines may elect to use this procedure. However, GE engines must be started individually, and Rolls-Royce engines have not been tested yet.

With a cold engine, the fuel comes on at about 21% of N_2 and starter cutout occurs at about 41% of N_2. Cashman said the auto-start system monitors EGT during the start sequence. If the first start attempt does not work, the automatic system will try a second and third

The 777 handles much like a traditional aircraft while providing pilots with a variety of protection and warning.

The 8-×-8-in. flat-panel, liquid crystal displays provide graphics and text in color and are easy to read in bright sunlight.

time using both igniters. It takes a long while for the big turbofans to spin up to idle, and these engines each took about 50 sec. N_2 peaked about 57% on both engines.

When we were cleared to taxi, I advanced the throttles. The 777 does not need a lot of thrust to break away, and after we began to roll, I pulled the throttles back to idle. There was plenty of power to carry us through a 180-deg. turn away from the hangar in front of us. The nose gear is 12 ft. behind the pilot's seat, so you have to delay the turn a bit if you want the wheels to track on the yellow line. The 777 has the same wing span as a 747, and the size felt comfortable to me based on my experience as a USAF Reserve C-5A copilot.

I turned down the taxiway and was initially a few feet left of centerline until I got properly oriented. The carbon brakes did not bind. Each main gear has six wheels, and when the pilot brakes below 30 kt. of ground speed, only four of the six brakes are used. The selection of brakes is automatically alternated, and I noticed no change in the braking action. We were taxiing at about 20 kt. based on the ground speed readout on the LCD display in front of me.

The 777 has two main landing gears with six wheels each. On landing, the touch-down sequence has no unusual characteristics.

As we taxied, I performed a flight control check. I did not feel much movement through the aircraft body when I deflected the rudder stop-to-stop, even though it weighs 1,500 lb. We completed our checklists and received clearance for takeoff on Runway 31L. I maneuvered the 777 onto the runway for a rolling takeoff. Winds were 135 deg. at 6 kt. (a slight tail wind) with a 4,400-ft. overcast ceiling.

I stood the throttles up to 1.05 EPR and let the thrust stabilize. I then depressed one of the takeoff/go-around (TOGA) buttons on the throttles and the autothrottle system advanced them to 1.364 EPR. The aircraft weighed 444,300 lb. as it rolled down the runway. A computer voice called out V_1 and Cashman said "rotate" as I pulled back on the control column and the aircraft rotated smoothly. We retracted the gear, and I picked up the flight director guidance, which called for a gradual pitch change to about 17 deg. Our clearance was to climb to 2,000 ft. on runway heading, expecting radar vectors after that.

Cashman explained earlier that the fly-by-wire system technically is in the direct mode (column to elevator) during takeoff. It then transitions to the C Star (C*) maneuver demand control law. In C* the fly-by-wire computer will deliver the pitch rate or "g" commanded by the pilot when he moves the control column. This is the case regardless of what else is happening as long as speed remains constant. Thus the computer may command elevator movement to compensate for wind, turbulence or a change in thrust without moving the control column. However, the 777 roll control surfaces are directed by wheel movement, rather than on a maneuver-demand basis.

The autothrottle system reduced power to a climb setting at 1,000 ft., and I heard a change in engine noise as the fan resonated. Cashman said the loudest engine noise seems to occur at this point, but noted that it is amplified a bit in the test airplane, which has no sidewall insulation. We retracted the flaps to 5 at 1,500 ft. and to a setting of 1 (slats only) at 2,000 ft. at about 180 kt. Passing 2,400 ft. at a speed of 200 kt., we brought the slats up and began accelerating to 250 kt. while climbing to 9,000 ft. The speed tape reminds the pilot of gear and flap placard speeds by marking the next restriction that applies with a red symbol.

When we were cleared by ATC above 10,000 ft., I followed the command bar to accelerate to the 320-kt. economy climb speed. This speed, usually 300-320 kt., is calculated by the flight management computer based on the aircraft gross weight and the cost index entered by the airline. Cashman had previously entered a cost index of 110, which is a typical value. Normal climb speed is close to V_{MO} (330 kt.). As we accelerated, I had to trim to adjust to the new speed.

When the pilot trims the aircraft, he is essentially changing the fly-by-wire computer reference speed, not moving the trimmable horizontal stabilizer. The computer controls the stab trim on its own. When the trim reference speed in the computer equals the speed of the aircraft, the forces on the pilot's control wheel are nulled. The control column forces buildup at about 3 lb. for every 10 kt. of airspeed change, so you can feel them if the aircraft is accelerating or decelerating and you are not trimming.

Craig S. Peterson, a 777 flight deck engineer responsible for fly-by-wire systems, said speed stability is one of the main differences between Boeing and Airbus approaches to fly-by-wire controls. Boeing is keeping the "feel" and control of the 777 similar to the rest of its fleet and giving pilots additional speed cues.

We accelerated slowly on our climbout, so it was easy to keep up with the trim. Our climb rate was about 2,000 ft./min. passing 12,500 ft. with a fuel flow of 16,900 lb./hr. on each engine.

Cashman pointed out that the navigation map display shows the situation behind the airplane as well as in front of it. The Honeywell FMS also keeps track of the four closest alternate airports in the navigation data base. It is not even necessary to call up the "legs" page on the FMS to go to one of the alternates. The pilot simply hits the "alternate," the "direct" and the "execute" buttons on the control display unit to begin a diversion. Data link will allow the company to let the pilot know when an alternate is below minimums and substitute another one with suitable weather. The FMS also provides an engine-out driftdown schedule.

All of this could be very useful to the crew of an ETOPS aircraft in an emergency far from land. The aircraft also has a highly redundant electrical system with backup generators on each engine to meet ETOPS requirements. All aircraft systems, including fuel, have been kept simple to operate.

We continued our climb while turning toward the North Training Area east of the Cascade Mountains. We planned to maneuver in that area between FL330 and FL370. Passing 20,000 ft., we were flying at 314 kt. (Mach 0.68), climbing at 1,500 ft./min. and burning about 14,000 lb./hr. on each engine.

We passed FL330 holding Mach 0.823, which we held to level off at FL350. At about 1,000 below our level-off altitude, we were burning about 11,000 lb. of fuel per hr. on each engine.

Cashman said the climb speed is very close to the initial cruise speed, so no trim adjustments are needed at the top of climb. We reached FL350 21 min. after brake release, and we then accelerated to

Mach 0.84. The wing was originally designed for a long-range cruise speed of Mach 0.83, but it has turned out to be more like Mach 0.8A, according to Cashman.

Once stabilized at Mach 0.84, I noted that the indicated airspeed was 288 kt., true airspeed was 480 kt., EPR was 1.104, N_1 was 80.4%, N_2 was 81.9%, EGT was 367C and fuel flow was 7,100 lb./hr. on each engine. After about 2 min. with the autothrottle system engaged, I checked these figures again. The readings were still very close. Our fuel flow was 7,000 lb./hr./engine and the airspeed/Mach was the same. Our gross weight was approximately 431,300 lb. at this point. Cashman said the autothrottle system will let the speed wander a bit to minimize throttle activity during cruise, and that is typical of modern transports.

The 777 fly-by-wire system uses three different modes for operation. Normal mode provides a variety of augmentation such as stall and bank-angle protection. A loss of air data and other malfunctions would result in the system reverting to the secondary mode. Most augmentation is lost along with the autopilot. Then there is the direct mode, which is the most degraded form of operation. The system would only go into this mode following serious malfunctions, which are highly improbable. The pilot also can select direct mode at any time using a guarded switch on the overhead panel, although no normal or emergency procedure calls for this.

None of the overspeed, stall and bank angle protection features are available in direct. Cashman selected the direct mode with the overhead switch so I could maneuver the aircraft briefly in this configuration. The 777 flies very much like a manual, cable-controlled airplane in this mode. There is no yaw damper, however, and the aircraft has a tendency to Dutch roll a bit, as I found when I moved the rudders. But other than that, the direct mode offers docile handling qualities for such a degraded mode. Few pilots will ever have to use direct mode in revenue service.

Cashman reselected normal mode, and I executed a 60-deg.-bank turn to the left. To maintain level flight, the pilot has to keep the wheel turned against the resistance provided by bank-angle protection and pull back on the column. But the pilot can still select any bank angle desired. I let go of the controls, and the fly-by-wire system quickly returned the aircraft to less than 30 deg. of bank. I then pulled back far enough on the control column during the turn to see the pitch limit indicator come into view on the attitude indicator. This symbol shows the pitch attitude at which stick shaker would activate if I continued to pull. It is displayed at slow speeds when the flaps are up and at all times when the flaps are down.

I got out of the seat, and another guest pilot did some maneuvering at altitude, then descended to 16,000 ft. for some low-speed evaluations. At this altitude, we did a "slam" acceleration on the left engine from idle. It took about 9-10 sec. for the engine to reach maximum continuous thrust. The aircraft does not have much of a tendency to pitch up when full power is applied in the normal mode because the fly-by-wire system compensates for thrust coupling associated with the large engines underneath the wings.

We began deploying flaps at 16,000 ft. after I returned to the left seat. Flaps 1 (slats only) were deployed on base leg during an approach, were selected at 195 kt. Flaps 5, which would be deployed on base leg during an approach, were selected at 195 kt. Flaps 20, which are normally deployed at about glideslope capture when the gear is extended, were selected at 178 kt. Few column inputs were needed during this process. The fly-by-wire system compensates with 2-3 deg. of nose-down pitch to maintain the some flight path as the flaps deploy. No trimming is required during extension and retraction unless the speed is changed. The same compensation occurs when speed brakes are deployed.

Buffeting was noticeable at Flaps 20. We set the speed bug over the numeral 20, which appeared on the speed tape at about 154 kt. This was the Flaps 20 maneuver speed, which means you can bank to 40 deg. and still not activate the stick shaker.

I throttled back to idle on the left engine and then began a turn in the direction of the "dead" engine. The fly-by-wire system assists the pilot in an engine-out situation by adding a little rudder when the control wheel is moved. Boeing is also developing a thrust asymmetry compensation (TAC) system that will take care of almost all of the yaw associated with an engine-out. This system will be flight tested this summer on the aircraft I flew. I worked on coordinating the turn by keeping the slip/skid indicator centered under the bank pointer on the electronic ADI. The asymmetry was not difficult to control.

I rolled out of the turn and throttled back the right engine a bit to slow to 133 kt. The fly-by-wire system would not let me trim the aircraft to a speed slower than this minimum maneuvering speed. Stall protection limits the trim reference speed at this point so that trim is inhibited in the nose-up direction. We were flying about 25 kt. above minimum control speed, which is below the 1-g stall speed. At this slow speed, large right pedal force was required to keep the aircraft from yawing. The dual-rate rudder trim switch can be used in the high-rate mode (2 deg./sec.) to quickly feed in rudder trim and compensate for the asymmetric thrust.

The gear, as well as the entire airplane, was designed on Dassault Systemes' CATIA software.

For the next maneuver, I advanced the left throttle to match the right one as Cashman hit the rudder trim cancel switch. This slowly removes rudder trim. We began a simulated go-around maneuver at Flaps 20 by advancing the throttles to TOGA. As I began a climb, Cashman cut the left engine to idle. The thrust decays gradually as the big fan winds down, and it is not difficult to keep up with the need for right rudder to maintain coordinated flight. Cashman said the engine even spools down slowly following a fuel cutoff. Boeing test pilots thought the loss of thrust would be a pretty dramatic event in the 777, but the inertia of the big fan makes it quite tame compared with other transport aircraft, according to Cashman.

With symmetric power again and both engines in idle, I began slowing to approach a stall with Flaps 20. We planned to go well beyond stick shaker and just short of a full stall. We started at 138 kt. at

16,000 ft. and began slowing at about 1 kt./sec. As we slowed below the top of the amber band on the speed tape (133 kt.), the nose-up trim stopped working as a stall protection. The auto slats also fired, meaning the leading-edge devices extended from sealed to gapped position to provide better handling qualities.

The stick shaker activated at 116 kt. We began to encounter buffet as I continued to pull to near full-aft column, and we came very near a 1-g stall. Cashman estimated I was pulling about 40-50 lb. of pressure when I released the column and advanced the power to recover just before actually stalling. If I had pulled the column all the way to the stop, it would have taken about 70 lb. of force.

Cashman said the 777 stall characteristics at Flaps 30 were originally "quite sporty," and flight test aircraft rolled as far as 110 deg. during tests in this configuration. Boeing engineers then tailored the fly-by-wire program so the outboard ailerons did not come down at high angle of attack and Flaps 30. The inboard leading-edge slots were also blocked a bit to produce less lift inboard. If the aircraft stalls during a turn, the fly-by-wire system commands a roll to wings-level.

I took another try at flying in the direct mode and found that the roll rate was slightly less than in normal mode at Flaps 20. Cashman said a stall in the direct mode would not be any different except that the force buildup on the control column would be quite a bit higher. Direct mode also lacks any compensation for thrust coupling, so the airplane also pitches up more when throttles are advanced to full power.

It was time to call Seattle Center for clearance to Grant County Airport to perform some touch-and goes there. The 8-×-8-in. map display made it easy to picture our route to the airport. We began retracting the flaps as I initiated a descent to 9,000 ft. We planned an ILS approach to Runway 32R at Grant County, where the winds were 360 deg. at 12 kt. and the ceiling was 1,300 ft. broken.

Cashman suggested that I try the speed brakes, and I extended the spoilers using the handle on the center pedestal. The buffeting did not seem too pronounced with flaps up, but it gets much more noticeable at Flaps 30, as I found later. Speed brakes cause less buffeting than on the Boeing 757 and 767, according to Cashman. I retracted the spoilers as I began leveling off at 3,000 ft., and we began extending the slats (Flaps 1) at 213 kt. Flaps 5 were extended next at 194 kt., and I slowed to 175 kt. or V_{REF} for 30 deg. of flap plus 40 kt. With the glideslope alive, we extended the gear and selected Flaps 20.

I was hand-flying the approach with autothrottles engaged, a technique that Boeing recommends for manual flight in the 777. The moving throttles made it easy to monitor power changes. Flaps 30 were selected when we captured the glideslope, and we settled into a well-stabilized approach at about 137 kt. (V_{REF} + 5 kt.) at a weight of 419,000 lb.

As the aircraft crossed the threshold of Runway 32R, the altitude callouts at 50, 30 and 10 ft. helped me judge when to begin my flare. As Cashman suggested, I waited until 20 ft. to begin a gradual flare. The result was a smooth touchdown. I lowered the nose, and Cashman reset the flaps and trim as we rolled down the runway. I advanced the throttles and, when the aircraft accelerated again, rotated the nose. We left the gear down and during climbout began retracting the flaps.

Cashman said later that the 777 is very easy to land, and this is partly due to ground effect. The airplane has a wing span as large as the 747, but the wing is closer to the ground on landing and ground effect is quite strong. Boeing engineers even put some ground-effect compensation (a slight pitch-down tendency) in the fly-by-wire control low for landing, but Cashman said it is hard to tell the difference flying with or without it.

The next approach was an autoland. We selected Flaps 5, slowed to 170 kt., engaged the autopilot and then captured the localizer. Soon we had intercepted the glideslope with gear down and Flaps 30, and the autopilot had us stabilized on a good approach path. Of course, the control column and throttles were moving to show the inputs being mode by the autopilot and autothrottle systems. These movements were not very pronounced compared with my memory of how earlier generation autoland systems used to jockey the throttles around.

By slewing the magenta box on the speed tape so it was just above the reference (REF) marker on the tape, I set the autothrottle to maintain V_{REF} + 5 kt. The spoilers were armed to deploy on landing, and the autopilot rounded out to a firm touchdown on centerline. I took over the controls as Cashman reset the trim and flaps. I then advanced the throttles and rotated the nose for takeoff. Cashman estimated that our rate of descent at touchdown was 3-3.5 ft./sec. New autoland software, which will reduce the descent rate by 1.5 ft./sec., should make touchdowns smoother.

Cashman pulled the right engine to idle just as I began to turn right to 120 deg. at 3,500 ft. with climb power set. I compensated with rudder pedal movement for a while before setting in rudder trim once we were stabilized on the downwind leg.

We planned a single-engine approach to a full stop at Flaps 20 with an approach speed of 143 kt. (V_{REF} + 5 kt.). We put the gear down and flaps to 20 at 170 kt. after intercepting the localizer and were soon established on the glideslope. The autothrottle controlled the left engine, and I could feel the single throttle move in my hand. Not much throttle movement was needed to keep the 777 on its approach speed. The 777 seemed easy to handle on an approach in a simulated engine-out condition at a gross weight of 413,000 lb.

Rudder trim was zeroed as I crossed the threshold, and the 50-, 30- and 10-ft. voice calls helped me time my flare. After another smooth touchdown, the spoilers deployed. Cashman suggested we go to maximum reverse on the left engine as a demonstration. We did not select thrust reverse immediately after touchdown, and it took a while for the engine to spin up. Directional control was not a problem. I applied the brakes at 70 kt., and we turned off onto a taxiway. In all, I had flown the aircraft for 1 hr. 45 min.

After the flight, Cashman said that Boeing 747-400 pilots will probably have the easiest time transitioning into the 777 (one foreign airline and one domestic carrier are planning to qualify some crews to fly both aircraft). Boeing 757/767 pilots also will find a lot that is familiar in the 777.

New handling equipment developed for PW4084

William B. Scott/Denver

United Airlines, as part of Boeing 777 certification, is demonstrating its mechanics are adequately trained to change a Pratt & Whitney PW4084, using unique support equipment built especially for handling the large powerplant at remote sites.

The demonstration for Federal Aviation Administration officials at United's Denver maintenance facility, which was scheduled for last weekend, is part of the airline's early-ETOPS clearance process. United's engine specialists must show they can repair a 777 powerplant at remote locations, using equipment they transport to the grounded aircraft site.

Mechanics that recently completed United's PW4084 engine maintenance training will use a special engine cradle, air/truck shipping stand, a collapsible fan dolly, a fan shipping frame and other equipment designed and built by Stanley Aviation Corp., a Cobham subsidiary company. The support equipment enables splitting the 120-in.-diameter, 16,700-lb. engine, then handling the fan case and core separately.

"Historically, the problem is in the core, not the fan, about 95% of the time," G.D. Lilja, Stanley's director of ground support equipment-product development, said. Some airlines operating the 777 will remove the PW4084 in the field—wherever the grounded aircraft happens to be—attach the fan to a new core, then remount the engine on the aircraft.

Because Boeing and Pratt & Whitney recognized that removing and repairing or transporting the 777's huge high-bypass powerplants would pose new problems for airlines, a "Big Engine Impact Team" was formed about three years ago. Representatives from Stanley Avi-

Stanley Aviation's shipping frame for the PW4084 engine fan section can be rotated from a vertical to horizontal position for transport in a C-130 cargo aircraft. The all-mechanical rotation system enables four technicians to easily handle the 119-in.-dia. fan in the field.

Continued

ation and Advanced Ground Systems Engineering—both experienced builders of ground-handling equipment—and customer airlines were invited to participate. Logistics factors identified by the group shaped design requirements for the engine-handling gear.

Stanley received a Pratt contract last summer to design and build four basic pieces of equipment in time for initial fit-checks in December. A process that normally would have taken up to 18 months had to be completed in about 90 days, according to Wim J. Lam, former vice president and chief engineer for Stanley. To meet this schedule, the company accelerated a planned switch to Pro/Engineer, a 3-D computer-based design system and instituted a concurrent design-manufacturing process. "We really had no choice," Lilja said. "It was the only way we could make the schedule."

A $250,000 investment in equipment, software and training enabled delivery of prototype equipment to Pratt last December. Initial fit-checks revealed the engine and equipment mated exceptionally well, requiring only minor modifications. The 3-D modeling tools precluded interferences that typically only show up during initial fit-checks, Lilja said. "The engine fit very well into the stands and fan frame. We were very pleased," William Kenney, an engineering staff representative for United, said.

One complex piece of equipment allows mechanics to attach an engine to a special cradle, then lower it onto a shipping stand and load the powerplant into a Boeing 747 or on a truck. The engine/stand combination had to fit through a 747 cargo door, which is 122.25-in. high. At its narrowest point—the 12 and 6 o'clock positions—the PW4084 fan has a 119-in. diameter.

Because the engine was so large, Stanley had to design equipment that would permit the powerplant to be shipped either as two pieces—the fan and core—or as a one-piece unit. An air/truck stand, built to carry the core, has integral shock absorbers for ground transport, but can be locked in place for airborne shipping on a standard pallet.

The fan removal and installation dolly and a fan shipping frame are unique to the 777's large PW4084. "These two pieces of equipment had never been designed before, as far as I know," Lilja said.

The collapsible dolly—used in separating the fan case from the core—is a steel and aluminum unit with 120-in. upright arms. Stanley's portable dolly supports the engine on a series of cables attached to those vertical arms, providing the flexibility of movement mechanics would have if changing the engine in a shop with an overhead suspension beam and winch. Manual cranks are used for repositioning the fan case in relation to the core. Integral dynamometers ensure loads also are balanced properly at each support point.

A separate fan shipping frame retains the fan case in either a vertical or horizontal position for transport on trucks or aircraft. The 120 in.-×-96-in.-×-140 in. frame rotates 90 deg., serving as a simple all-mechanical "rollover" device that can be operated by four technicians. The fan case can stand vertically in a 747 aircraft or on a truck; it must ride horizontally in a C-130. An integral "fan blade box" located in the center of the frame can hold 22 of the 3-ft. long, 40-lb. fan blades.

Stanley is developing additional shipping and lightweight work stands, an engine-handling sling and other specialized equipment to accommodate 777 operators' particular maintenance needs.

Certification tests continue on GE and Rolls-powered 777s

Paul Proctor/Seattle

Unprecedented joint FAA and European type certification in mid-April of the new Boeing 777 transport capped an intensive six-aircraft, year-long flight test program that totaled more than 1,600 flights and 3,000 flight hours.

The certifications, which cover a total of 19 countries as well as FAA production authority, came almost exactly five years after a formal program announcement by Boeing. FAA certification was received the same week it was originally targeted, according to Gerald Mack, Boeing director of certification and government.

All aircraft performance guarantees were met or exceeded, according to Boeing. The 777 seats 305-328 passengers in a three-class configuration. Boeing holds firm orders for 144 of the aircraft from 15 airlines.

The 777 flight test program, based at Boeing Field near downtown Seattle, currently is performing trials leading to certification of GE-90- and Rolls-Royce Trent 800-powered 777s. The first GE-90-powered 777 flew in February and is scheduled for delivery to British Airways in September, 1995.

The first Trent 800 engine now is being installed on a 777 for Hong Kong's Cathay Pacific Airways and will fly by the end of May. Boeing's flying testbed 747, the first 747 built, recently completed performance and operability testing of a Trent 800 mounted on one of its four engine pods. The configuration simulated 777 use.

Boeing believes it is on schedule for the mid-May, entry-into-service award by FAA of extended-range twin-engine operations (ETOPS) authority up to the maximum 180-min. from an alternate airport. Serial No. WA004 is performing those tests and currently is on a 90-cycle pre-service tour with United Airlines.

First 777 transport for Japan's All Nippon Airlines is shown during final body join at Boeing's Everett, Wash., wide-body manufacturing center.

Top FAA officials at the 777 certification ceremony here in mid-April said the 777 appears on-track to meet Boeing's mid-May ETOPS deadline. Boeing also is spooling up testing for the longer-range B-market version of the 777. Although the first B-market 777 will not fly until 1997, certain of these tests, such as braking performance, can be satisfied using existing A-market aircraft at heavier weights.

By first delivery to United Airlines this month, the Pratt-powered aircraft will have accumulated slightly over 3,300 flight hours and 4,000 ground test hours, according to John Cashman, chief pilot, 777 Division, Boeing Commercial Airplane Group.

The flight test totals are roughly double those of the 757 and 767, Boeing's most recent new-aircraft certifications. Both the 757 and 767 flew about 1,700 hr. of certification testing each in the early 1980s, Cashman said. WAOOI, which primarily performed aerodynamic performance, stability and control tests, flew its certification schedule at about a 75-hr./mo. rate, Cashman said.

Key to the nearly trouble-free flight test program was the intensive ground test and engineering simulation of aircraft components and systems performed in Boeing's Integrated Aircraft Systems Laboratory prior to the first takeoff (AW&ST Apr. 11, 1994, p. 56).

Early customer signoff on aircraft configuration virtually eliminated the unplanned design changes, which impeded previous flight test programs, Cashman said. Overall, the 777 flight test program will involve a total of nine aircraft, 4,800 cycles and 6,700 flight hours. It is scheduled to be completed in March, 1996.

BOEING 777 SPECIFICATIONS*

POWERPLANTS

Pratt & Whitney PW4077, 77,200-lb.-thrust rating; General Electric GE90-76B, 76,400-lb.-thrust rating; and Rolls-Royce Trent 877, 76,900-lb.-thrust rating. Note: WA--1 has flown tests with PW4077 at 77,200 lb. rating and other tests with PW4084 at 84,00 lb. rating for B Market aircraft. Rating is changed by a software modification to electronic engine control.

WEIGHTS

Maximum takeoff weight	535,000 lb. (242,671 kg.)
Maximum landing weight	445,000 lb. (201,848 kg.)
Typical operating weight empty	302,000 lb. (136,984 kg.)
Maximum payload	118,000 lb. (53,524 kg.)
Maximum fuel capacity	207,700 lb. (31,000 gal.)

DIMENSIONS

Length	209.1 ft. (63.7 meters)
Fuselage exterior diameter	20.3 ft. (6.2 meters)
Tail height	60.8 ft. (18.5 meters)
Wing span	199.9 ft. (60.9 meters)
Wing area	4,605 sq. ft. (427.8 sq. meters)
Wing sweep	31.6 deg. at 1/4 chord
Fan diameter	PW4077 - 112 in. (2.84 meters) GE90 - 123 in. (3.12 meters) Trent 877 - 110 in. (2.79 meters)

PERFORMANCE

Takeoff field requirement, sea level ISA +15 (typical mission)	5,500 ft. for 1,000-n.m. leg
Maximum range 375/400 passengers	PW4077 - 4,615 n.m./4,390 n.m. GE90-76B - 4,595 n.m./4,370 n.m. Trent 877 - 4,695 n.m./4,470 n.m.
Maximum operating speed	Mach 0.87
Long range cruise	Mach 0.84
Roll rate	Approximately 20 deg./sec.
Fuel jettison rate	5,200 lb./min.
Typical passenger loading (2 class)	375 (9 abreast); 400 (10 abreast)
Under floor capacity	32 LD3s, 600 cu. ft. bulk (total volume = 5,656 cu. ft.) *or* 10 pallets, 600 cu. ft. bulk (total volume = 4,750 cu. ft.)

*For A Market 777

Airbus' new twins surprisingly similar

David Hughes/Toulouse, France
October 4, 1993

Transition to A330 side-stick controller is easier than expected for a former heavy transport pilot.

Flying the Airbus A330 and the A321 on the same day provided a good opportunity to see the similarities and differences between the two aircraft. The cockpit layouts are almost identical, and the tuning of the normal control law on the fly-by-wire flight controls makes the handling qualities very similar, despite the fact that the A330 is a wide-body and the A321 a narrow-body aircraft. Highly standardized display formats and system architectures will help make it possible for a pilot to be qualified to fly both aircraft at the some time.

Airbus Industrie has succeeded in creating two new members of its fly-by-wire transport family that handle surprisingly like one another, even though the A330 can weigh 284,000 lb. (128,800 kg.) more than the A321 at takeoff.

This *Aviation Week & Space Technology* pilot flew both aircraft here recently. From 1973 to 1984, I flew heavy transports, mostly in the Lockheed C-5A with the U.S. Air Force Reserve. This was my first exposure to Airbus' computerized fly-by-wire flight controls and side-stick controllers.

Fernando Alonso, deputy director of the Test and Development Dept., said the ambitious A330 and A321 flight test programs have been largely trouble-free. However, some additional requirements have emerged. About 50 flying hours were added, for example, to adapt control laws to match Dutch roll characteristics in the A321.

A330 rotating main bogey touches down in two-step sequence followed by the nose gear during many touch-and-go landings.

The Dutch roll turned out to be less divergent in flight tests than expected, prompting the change.

Our group received a thorough preflight briefing from chief test pilot Pierre Baud, vice president of the Airbus Flight Div. We left with Baud to board the General Electric CF6-80E1-powered A330. Like the much smaller A321, the A330 cockpit is dominated by six 7.25 × 7.25 in. cathode ray tube displays. Each pilot has two displays, and there are two Electronic Centralized Aircraft Monitor (ECAM) displays. Both cockpits employ the some lights-out design.

It was easy for Baud, who took the right seat, to set up the Honeywell Flight Management System for departure by entering a few key parameters such as our ramp weight of 370,372 lb. (168,000 kg.) and our fuel weight of 86,841 lb. (39,391 kg.). Our center of gravity was 26%, and our reference speeds for takeoff were 131 kt. for V_1 and V_R and 136 kt. for V_2. These were displayed on the pilot's FMS control/display.

We pushed back and after being disconnected from the tug, Baud started the No. 2 engine. After setting the engine master switch to auto start, the entire start sequence progressed automatically. The start valve closed automatically and EGT peaked at about 450C before falling back to 370C. N_2 stabilized at about 22%. The No. 1 engine then was started. One glance at the ECAM display showed that all checklist items had been completed. "Finished," Baud said.

We were soon cleared to taxi down the runway for a 180-deg. turn, during which Baud added outside throttle to bring us around. The guest pilot in the left seat applied the brakes and advanced the throttle to 50% of N_1 on the two 67,500-lb.-thrust General Electric engines. Baud wanted to see that both of the large turbofans were accelerating normally before selecting full power.

The pilot in the left seat then released the brakes and moved the throttles forward through two detents to maximum continuous/flex thrust. We were making a reduced thrust takeoff rather than selecting maximum takeoff/go-around in the third detent. Acceleration was smooth. We rotated at 131 kt. through 10 deg. and finally to about 15 deg. of pitch. Baud had said a pitch attitude of 15-17 deg. would be appropriate at this weight, slightly higher than the 13 deg. we would see later in the A321.

The pilot retracted the gear and followed the flight director commands. As we passed through about 1,500 ft. agl., a flashing "throttle lever" message was displayed, which cued the pilot to reduce thrust. The throttles were moved back one detent into climb and the autothrust system automatically engaged.

Baud explained later that only one climb thrust setting is being used on this A330 in the test program, but two or three settings, including derated ones, will be available in the production aircraft. A pilot will be able to select on the FMS the desired climb power setting.

At 3,000 ft., the nose was lowered farther so the aircraft could accelerate to 250 kt., and the flaps were retracted after we passed "F" speed. An "S" appeared next on the speed tape at about 177 kt., showing us the minimum slat retraction speed, and Baud brought in the slats.

At 6,000 ft., the fuel flow to each engine was 18,298 lb. per hr. (8,300 kg.). We were climbing at about 3,000 ft. per min. and sustained this rate through 10,000 ft., where we accelerated to the optimum climb speed of 302 kt., as shown on the flight director command bar. As we passed through 12,000 ft., the pilot in the left seat noted how low the workload was by saying, "There has got to be more to do with your hands." Baud took this as a cue to show us the writing table that slides out from under the instrument panel in front of each pilot.

Fuel flow was down to 14,440 lb. (6,550 kg.) per engine as we passed through 20,000 ft. Next the maximum 67 deg. of bank the normal control law will allow was demonstrated. Pitch trim is automatic through 33 deg. of bank but cuts out after that. The bank angle looked even steeper than 67 deg., but it is well beyond what an airline pilot would select.

At 30,000 ft., fuel flow was down to 10,028 lb. (4,550 kg.) per engine, and we were climbing at about 2,000 ft. per min. We planned to accelerate to Mach 0.82 at 33,000 ft. but got into enough turbulence in the clouds that Baud requested a higher level. The aircraft handled the turbulence well, and we were soon cruising smoothly above the clouds at 37,000 ft. with a fuel flow of 8,113 lb. per hr. (3,680 kg.). It was 20 min. since brake release.

Next Baud simulated a hydraulic system failure to show that the ECAM handled it in the A330 in the same way that it would in the A321. The display formats were the same, even though the hydraulic systems and related messages were different in some ways.

A seat change was made at 37,000 ft. The second guest pilot evaluated the A330's performance in the fly-by-wire system's normal control low by pulling full back stick in an evasive maneuver to the right. Indicated airspeed decreased to 192 kt. as we reached 41,000 ft.

Gilles Robert, the flight test engineer on the aircraft, then used a test program computer to select the flight control system's direct law. The chances that a line pilot will ever see this backup mode are slim because the combination of failures that would initiate it have a 10^{-7} probability. In direct law, none of the speed protection features are available, and the aircraft is limited to Mach 0.80 and 330 kt. Pitch trim reverts to manual and the yaw damper remains, but maneuvering must be done with care.

I took the left seat while we were still at 33,000 ft. and Mach 0.80. Back in normal law, I got a chance to see how the side-stick controller allows you to bank up to 33 deg. and release the stick to have the control law feed in elevator to compensate automatically. Beyond 33 deg. this feature cuts out and you must feed in back pressure to maintain altitude. I used the arm rest with the controller but learned later that many experienced A320 pilots dispense with it except in turbulence.

The next maneuver was to initiate a descent to 12,000 ft. and evaluate the V_{MO} speed limiting protection. As I pushed the nose over, the speed eventually peaked at 345 kt. Speed protection features would

Air Inter is first customer for the General Electric CF6-80E1-powered A330 and plans to use the aircraft in domestic service.

not allow us to exceed V_{MO} plus 10-15 kt. Baud then demonstrated the speed brake by deploying it half way at first and then fully. There was some buffeting, but the ride was comfortable and we reached a descent rate of about 5,000 ft. per min. Rates of 6,500 ft. per min. can be obtained in a constant Mach number descent.

Baud retracted the speed brakes and instructed me to push full stick forward, which I did. The normal law limits nose down pitch to 15 deg. When I released the stick, the aircraft recovered on its own as the fly-by-wire system raised the nose and we recorded 1.7g. Our deceleration had been about 5-8 kt. per sec. I next established a 7.5 deg., nose-low attitude and then pushed the stick full forward until the speed tape penetrated the red barber pole region denoting V_{MO} (330 kts.). At about 8 kt. above V_{MO}, I released the stick for the automatic pull up. We were decelerating at about 1 kt. per sec. and therefore recorded only 1.259 before recovering. The protection system will not let the aircraft exceed 2.5g or 2g with slats and flaps extended.

At 12,000 ft. we then began a low-speed evaluation with the throttle back as I maintained level flight and the aircraft decelerated in a turn. Low-speed protection symbols soon came into view on the speed tape, with the first being a thin line denoting the beginning of the lowest selectable speed (V_{LS}) or about 1.23 V_S (stall). An airline pilot would not normally fly below this speed, and a "Speed, speed" voice warning notified us we were entering a low energy situation.

I continued to decelerate until a thicker barber pole appeared, denoting the beginning of the alpha protection region. The normal law shifts into an alpha control mode at this point, and Baud said I should apply full back stick until the alpha floor feature was triggered. Alpha floor is not marked on the speed tape because it varies with the deceleration rate.

At alpha floor, the autothrust system selects takeoff/go-around power. With the stick still in the full back position and the pitch about 20 deg. nose high, the aircraft was sitting safely at the maximum allowable angle of attack (AOA). In a clean configuration we reached 12 deg. AOA, which was still about 1-1.5 deg. less than C_{Lmax}. As the aircraft climbed in this configuration, I was able to comfortably turn 30-deg. left and then reverse course and turn 30 deg. right with full back stick.

As a pilot who had never flown an Airbus aircraft before, I found maneuvering a large transport at full back stick and maximum angle of attack to be a remarkable experience. A feature that allows a line pilot to quickly obtain the maximum short-term performance of a heavy aircraft without penetrating the stall region or over-stressing the airframe is an invaluable tool.

Following this maneuver, Baud extended the landing gear and full flaps, and I pulled full back stick again, decelerating a little more rapidly. The speed dropped to a little less than 100 kt. as the aircraft reached a maximum 15-17 deg. AOA. The pitch attitude felt quite steep, but the aircraft was easily maneuverable at full back stick.

Even though the throttles were in the idle detent, the autothrust system was delivering full power. Baud next demonstrated a similar maneuver with an even more abrupt entry that the fly-by-wire system handled smoothly.

Since it was time to set up for the approach to Runway 15R at Toulouse, we began an idle descent toward the airfield while discussing landing procedures. When Baud activated the approach mode on the FMS control display unit, the system calculated all of the speed targets and presented them on the speed tape.

It was a bit disconcerting to have the throttles in a fixed position during approach. I flew with my left hand, leaving my right hand free. The autothrottle system controlled the power. Clearly the fixed throttle position would take some getting used to, and a pilot needs to keep N_1 values in his crosscheck.

I was soon stabilized on the ILS glideslope and localizer below the clouds with winds calm, while computer voice calls keyed to the radio altimeter began coming at 100 ft. intervals below 500 ft. At 50 ft. the pace picked up to calls every 10 ft., and at 40 ft. I applied a little back stick to flare. At 20 ft. a voice command of "retard, retard" prompted Baud to pull the throttle to idle. The rear tires on the two main gear touched down smoothly, and I left the stick alone until the front wheels on the "rocking" bogies made contact. This was easy to feel and hear. Then I applied just a little back stick to reduce the pitch down, and the nose wheel also met the runway smoothly. Other than being a little left of centerline, the landing had worked out quite well.

Baud reset the flaps and slats, stood the throttles up and then selected takeoff power, leading up to a call of "rotate." I pulled back on the stick and headed for 15 deg. and the nose came up smoothly. The side stick was feeling more comfortable all the time, and I turned right and climbed to 2,000 ft. for another circuit. We practiced engine-out control techniques on downwind, and I had difficulty interpreting the electronic turn and slip indicator at first.

The next ILS approach went well and the touchdown was pretty much the same as my first one. Right after liftoff from the touch and go, Baud retarded the throttle for the right engine and I initially lowered the pitch attitude a few degrees from 15 deg. as instructed. The next move was to apply left rudder to compensate for the uneven power situation, and the aircraft climbed nicely in the simulated engine-out configuration.

This maneuver concluded my time in the A330's left seat. We then performed two more touch-and-go landings, followed by a full stop. Deceleration on the full stop was smooth with application of reverse thrust and ground spoilers. We then took off and climbed to 15,000 ft. with the third guest pilot in the left seat.

At 15,000 ft. in direct law, a Dutch roll maneuver was induced. The yaw damper immediately stopped it. Baud took this opportunity to show us an actual stall, which was possible with speed protection removed. We felt an unmistakable buffet and a pitch down motion in the clean configuration stall at 16 deg. AOA. Then with full flaps out, Baud slowed to stall the aircraft at 95 kt. and about 25 deg. AOA. The buffet was stronger this time.

We did not use speed brakes during the descent to the airfield, as this is not allowed in direct law. On one touch and go, the main gear touched down as before, but the nose gear made firmer contact. More stick movement is needed in direct law than in normal law to move the fairly small elevator.

During the next touch and go landing in normal law, the pilot depressed the auto thrust disconnect button on the throttles before beginning to retard them slowly to idle in the flare. The correct A330 procedure is to pull the throttles directly to idle without depressing this button. The throttle went into manual mode and the thrust began to advance to match the still fairly high throttle position—an unexpected development for those of us new to the airplane. Baud elected to go around. He said later that a complete explanation of the throttle system in the airline training course would help pilots avoid this situation. The final landing was uneventful and thrust reverse was applied. We had logged a total of 3 hr. of flight time.

It was interesting to see that a pilot with experience in round-dial cockpits, like myself, can become comfortable quickly in Airbus fly-by-wire transports. Jean Pinet, president of Airbus' Aeroformation training organization, said 80% of the pilots coming to Airbus have no glass cockpit or side-stick experience, so my background is relevant. By the end of the training course these pilots will surely be comfortable in the aircraft. However, A330 pilots will continue to climb a learning curve as they master less-critical aspects of the automated system.

Airbus currently has 127 A330s on order, but some deliveries could be delayed by leasing companies. The General Electric-powered A330 is slated for Joint Airworthiness Authority type certification this month to be followed in April by the Pratt & Whitney-powered version and in November, 1994, by the Rolls-Royce-powered aircraft.

A321 flight demonstrates design links to A330

David Hughes/Toulouse, France

Flying the Airbus A321 right after completing an evaluation of the A330 provided an opportunity to appreciate how cross training in aircraft differences will allow pilots to qualify on both aircraft. The narrow-body A321 is slightly more agile than the larger A330 because of the difference in roll inertia. However, Airbus has managed to achieve overall equality in the two aircraft as far as handling qualities are concerned due to careful "tuning" in the fly-by-wire control system.

It is not surprising, in light of this, that Airbus' Aeroformation training subsidiary proposes to train A320/321 pilots to fly the A330 in cross crew qualification courses lasting just 13 working days rather than 25 days or more.

The 186-passenger A321 has two fuselage plugs in it that extend the length of the fuselage by 273 in. (6.93 meters) over that of the 150-seat A320. The A321 has also been fitted with double-slotted flaps.

I flew the A321 with test pilot Nick Warner and two other guest pilots. After flying the A330, entering the A321 cockpit gave an immediate impression of having been there before, except that the cockpit was smaller.

I took the left seat, and Warner discussed the glare shield-mounted Flight Control Unit that allows the pilot to inject speed, heading and altitude changes to accommodate air traffic control (ATC) changes without having to go head-down to load them into the flight management system (FMS) control/display. The A321 FCU has an expedite button that is not found on the A330. This commands the speed for best lift/drag, or about 1.4 V_s.

We were soon pushed back and prepared for engine start. We fired up each of the CFM56-5B 30,000-lb. thrust turbofans just by moving an engine mode selection switch to start and placing the master switch to on.

The automatic start sequence was similar to that used in the A330 with its 67,500-lb. thrust GE engines. The exhaust gas temperature peaked at about 450C and dropped to 430C, and N1 stabilized at 20%.

The winds were calm, and Warner requested clearance to take off on Runway 33L. We had 30,758 lb. (13,952 kg.) of fuel on board and weighed 160,850 lb. (72,961 kg.) on the ramp. Warner said our center of gravity at takeoff would be 23%, and we would probably burn 440 lb. (200 kg.) taxiing out. The outside temperature was 22C, and our takeoff speeds would be 134 kt. for V_1; 144 kt., V_r, and 145 kt., V_2. These were close to the A330 speeds.

I released the parking brake and grasped the inner hub of the nose wheel steering tiller instead of the top of the arm. The steering is considered light to the touch, so this is the best grip. It was easy to control the A321 using the electrical nose wheel steering in this fashion.

I taxied onto Runway 33L and began a rolling takeoff by advancing the throttles to 50% of N1 before selecting flex takeoff/maximum continuous thrust at the second throttle detent. The aircraft accelerated smoothly as I directed it with rudder pedal steering. We accelerated to rotation speed, and I initially pulled the spring-loaded side stick back about halfway to initiate a pitch-up to 13 deg.

We lifted off about 5,000 ft. (1,500 meters) down the 11,482-ft. (3,500 meter) runway, and Warner selected gear up. At this point the flight director's speed reference system (SRS) commanded a pitch rate to maintain about V_2 plus 10. At 1,500 ft. above the ground, a flashing "throttle lever" message cued me to move the throttles back to the climb detent and engage auto thrust. At 3,000 ft. above the ground we began accelerating for the rest of our climbout. Warner selected configuration 1 (flaps up) followed in awhile by slats up. As we continued the climbout at 250 kt. we experimented with the expedite function, which commanded a pitch change to capture a green dot marker on the speed tape to accelerate our climb as if to meet an ATC request. We soon reached about 4,500 ft. per min. of vertical velocity.

Passing through 15,000 ft. the fuel flow was 7,319 lb. per hr. (3320 kg.) for each engine, which decreased to 6,349 lb. (2,800 kg.) by 20,000 ft. We were cleared to 31,000 ft., but I asked Warner what would happen if we were given an altitude restriction at 29,000 ft. en route. The flight test FMS software (not the final production version) would not take the change because it seemed too close to our top of climb. FMS and ATC requirements may not always be in synch in automated aircraft.

It was easier to master the touchdown sequence in the A321 in a series of touch-and-go landings than in the A330. The A321 is also a bit more agile in roll.

Passing through 25,000 ft. each engine was consuming 5,511 lb. per hr. (2,500 kg.), and we were climbing at 1,500 ft. per min. We passed 29,000 ft. at Mach 0.77, climbing at 1,100 ft. per min. and reached 31,000 ft. 17 min. after brake release. Each engine was consuming fuel at 3,527 lb. per hr. (1,600 kg.) at this point. Typical cruise values are 1,200-1,300 kg. per engine (2,645-2,865 lb.).

At 31,000 ft. we accelerated to Mach 0.80, and I initiated a 45-deg.-bank turn and began to add back pressure until feeling the onset of Mach buffet at 1.7g. Warner said later that it is possible for the pilot to pull the stick full back in the Mach speed range to achieve a maximum angle of attack without heavy buffeting.

One of the other pilots took over and flew the aircraft at altitude in direct law before giving up the seat to the third guest pilot. We were back in the normal law for a descent at this point. Warner selected a "managed speed" with a speed target of 250 kt.

As we continued the descent, however, Warner selected a speed of 300 kt., and we accelerated until the "barber pole" on the speed tape signifying the V_{mo} region came into view. The autopilot was disconnected passing through 25,000 ft., and the speed brake was deployed. I noticed the vertical velocity peaking at 8,400 ft. per min.

With the speed brakes back in, we kept a 5-deg. nose low attitude with the power up and got an overspeed warning (a beeping sound) as the normal control law started to feed in back pressure on the stick. With a 2.5 deg. nose down attitude at 360 kt., we were still protected from overspeeding the aircraft. Warner noted the system prevents a "massive exceedance." We also looked at a 5 deg. nose down attitude with 40 deg. of bank heading back into the V_{mo} band region. When the stick was released, the normal law rolled wings level while applying back pressure to effect a recovery.

After some other maneuvers, I was invited back into the seat to explore the low-speed safety features between 8,000 and 14,000 ft. The first low-speed maneuver involved slowing down with flaps in configuration 2: gear up and the Alpha floor feature disconnected to slow down the sequence of events for easier observation.

The pitch attitude reached about 15 deg. nose high soon after I pulled the stick full back, then slowly tapered off back toward 10 deg. The normal control law would not let me reach more than 15 deg. angle of attack, so I could not stall the aircraft.

The next low-speed maneuver was performed at flaps configuration 3, gear down and with Alpha floor engaged. With the power in idle I maintained the pitch attitude at 10 deg. as we slowed down and angle of attack rose above 10 deg. When we were 12 kt. below V_{LS} a "speed, speed" voice warning sounded. Then just as Alpha floor was

After takeoff, the A321 climbs at V_2 plus 10 kt. to 1,500 ft. before the throttle is moved to the climb detent to engage auto thrust.

triggered and takeoff/go-around power selected, I abruptly pulled the stick full back and reached 25 deg. of pitch attitude in just a few seconds. The angle of attack surged to 15.6 deg. before falling back to 13 deg.

I noticed we were climbing at about 2,300 ft. per min., and once again the aircraft was perched safely at the maximum angle of attack at a comfortable margin below stall. It was no problem to control the aircraft in this configuration.

Following a seat change another of the guest pilots began a descent toward the airfield with speeds being set by Warner. A typical descent schedule calls for deceleration from 250 kt. clean to approximately 210 kt. (green dot speed) at about 5,000 ft. Once we slowed another 30 kt. to "S" (slats) speed on the tape, the slats could be deployed. We were flying with a somewhat nose high pitch attitude when configuration 2 was selected by Warner as we passed below "F" (maximum flaps extension speed). This flattened the aircraft attitude quite a bit, to 2.5 deg. nose up.

At this point we decelerated toward our approach speed of 132 kt. After glideslope interception the gear was extended and configuration 3 was selected. The guest pilot flying made two good touch and goes, and it was my turn to get in the seat again. As with the A330, I found it easy to stabilize the A321 on the ILS with the fly-by-wire control system.

The same sequence of voice calls as heard in the A330 sounded, starting at 500 ft. and increasing to one call every 10 ft. beginning at

50 ft. As the aircraft passed through 40 ft., I put in a little back stick to flare. When the "retard, retard" voice call sounded I pulled the throttles quickly to idle. Once the main gear touched down, it only took a little back stick to soften the nose wheel contact. The touch-down sequence is much easier to master in the A321.

After we took off again and turned out to enter the downwind leg, there was enough time to observe a simulated failure of two out of three air data units. The fly-by-wire flight control system shifted from normal to alternate law, making the roll axis a little more lively. The stick-to-roll-surface relationship becomes direct in this case.

We were back in normal law by the time we were cleared to turn to base from our extended downwind leg. The next touch and go went as well as the first one and on the takeoff Warner retarded one of the throttles to simulate an engine out. It was easy to control the asymmetric thrust with rudder.

A final seat change was made, and the other guest pilot climbed out to a medium altitude to examine direct law control once again. The final two touch and goes were also good ones followed by a simulated engine out on the second takeoff. The full stop was flown with a simulated engine out to a smooth touchdown with maximum reverse thrust being applied down to a speed of 70 kt. We had flown 2.5 hr. and used 20,475 lb. (9,290 kg.) of fuel.

I felt even more comfortable now with the fly-by-wire control system and could see transitioning from a yoke to a side stick is not that difficult. It still did seem a bit challenging, however, to keep track of the power settings without any tactile feedback from the throttle quadrant.

Airbus has 153 A321s on order, and there are already more than 400 A320s operating in the world and an additional 230 on order. The IAE-powered A321 is slated for Joint Airworthiness Authority certification in December and the CFM-powered version in February, 1994.

AIRBUS INDUSTRIE A330 AND A321 SPECIFICATIONS

A330

POWERPLANTS

GE CF6-80E1, 67,500 lb. thrust rating; PW4000, 64,000 lb. thrust rating basic, 68,000 lb. optional; Rolls-Royce Trent 700, 67,500 lb. thrust rating basic, 71,100 lb. optional

WEIGHTS

Maximum takeoff weight	467,400 lbs. (212 tonnes)
Maximum landing weight	383,600 lbs. (174 tonnes)
Typical operating weight empty	264,300 (119.9 tonnes)

Maximum payload	97,300 lbs. (44.1 tonnes)
Maximum fuel capacity	171,855 lbs. (25,650 gal.)

DIMENSIONS

Length	209 ft. (63.7 meters)
Fuselage diameter	18.5 ft. (5.64 meters)
Overall height	55.2 ft. (16.8 meters)
Wingspan	197.8 ft. (60.3 meters)
Wing area	3,890 sq. ft. (362 sq. meters)
Wing sweep	30 deg.

PERFORMANCE

Takeoff field requirement, sea level, ISA + 15 (typical mission)	5,800 ft. (2,000 n.m. leg)
Maximum range	4,500 n.m. (355 passengers + baggage)
Maximum operating mach number	0.86
Typical passenger loading, (2-class)	335
Under floor capacity	32-33 LD3s or 11 pallets Also, 695 cu. ft. bulk

A321

POWERPLANTS

CFM 56-5B, 30,000 lb. thrust rating basic, 31,000 lb. thrust rating optional; IAE V2530, 30,000 lb. thrust rating

WEIGHTS

Maximum takeoff weight	183,000 lbs. (83 tonnes)
Maximum landing weight	162,000 lbs. (73.5 tonnes)
Typical operating weight empty	105,500 lbs. (47.9 tonnes)
Maximum payload	47,700 lbs. (21.6 tonnes)
Maximum fuel capacity	41,942 lbs. (6,260 gal.)

DIMENSIONS

Length	146 ft. (44.5 meters)
Fuselage diameter	13 ft. (3.96 meters)
Overall height	38.7 ft. (11.8 meters)
Wingspan	111.8 ft. (34.1 meters)
Wing area	1,320 sq. ft. (123 sq. meters)
Wing sweep	25 deg.

PERFORMANCE

Takeoff field requirement, sea level, ISA + 15 (typical mission)	6,000 ft. (1,000 n.m. leg)
Maximum range	2,300 n.m. (185 passengers + baggage)
Maximum operating mach number	0.82
Typical passenger loading, (2-class)	185
Under floor capacity	10 LD3-46(w)s + 208 cu. ft. bulk or 1,828 cu. ft. bulk

A340 handling, cockpit design improve on predecessor A320

David M. North/Toulouse, France
July 13, 1992

Airbus Industrie is offering the long-range, four-engine A340 with the same high-technology standards and almost identical handling characteristics as the smaller, twin-engine A320. The A340 is the second in the Airbus series of transports equipped with a fly-by-wire flight control system, side-stick controller and digital avionics. The A330—third in the series—has the same airframe as the A340-300 but is powered by two engines. It is scheduled to fly later this year.

A flight in the No. 4 prototype A340 by this *Aviation Week & Space Technology* pilot in late June convinced me that Airbus has nearly duplicated, and in some cases improved on, the excellent flying qualities of the smaller A320. The flight was from the primary Airbus facility here, and it was again with Pierre Baud, vice president of the consortium's flight division. I flew the A320 with Baud in 1987 (AW&ST Nov. 30, 1987, p. 40). Four pilots evaluated the A340 on this flight. The other three were a USAir McDonnell Douglass MD-80 captain, a retired British Airways Lockheed L-1011 captain and another journalist with varied piloting background.

During the past few years, I have flown a number of smaller aircraft from trainers to fighters, but have not logged time on large transports. The consensus among the four of us was that the A340 handled much like the A320 and gave the impression of being a smaller aircraft. This was especially impressive when considering that the A340 is a 558,900-lb. maximum takeoff weight aircraft and the A320 has a 162,000-lb. maximum takeoff weight.

Airbus Industrie A340 is powered by four CFM-56-5C engines. The aircraft is scheduled to receive European certification in mid-December and U.S. approval the following month.

Commonality between the A340 and A320 is an important element in the Airbus plan to develop a family of aircraft. The two transports have similar flight characteristics, cockpit layout and presentation, as well as normal, abnormal and emergency procedures. It is anticipated that pilots flying the A320 will have cross crew qualifications with both the A340 and the A330. The U.S. Federal Aviation Administration has already started its evaluation of the pilot qualifications and level of difference training needed to operate both the A320 and A340.

The ability to transition from the A320 to the A340 with a minimum of training to acquire separate type ratings is especially important to Northwest Airlines. Northwest operates A320s and is scheduled to receive the first of its 24 ordered A340s in March, 1993. Airbus officials anticipate FAA cross crew qualification approval in October. The company expects that it would take 13 working days to fully transition an A320 pilot to A340 qualifications.

Following a preflight briefing, we boarded the A340-200. The No. 4 aircraft is involved with performance and system testing and is still equipped with flight test instrumentation. The four aircraft now flying have accumulated more than 840 flight hours and 420 landings. The goal is for 2,000 flight hours. European certification is scheduled for mid-December and U.S. certification in January, 1993.

Gerard Guyot, Airbus test and development director, occupied the test station in the aircraft and was able to supply performance data during the entire flight. Because of the unfinished interior and the lack of some soundproofing, Baud said the cockpit noise would be slightly higher than that in the finished A340s.

I occupied the left seat while Baud took the right seat. The A340's zero fuel weight at the ramp was 284,760 lb. (129,200 kg.), including flight test equipment and water ballast. The fuel on board totaled 130,000 lb. (59,000 kg.), or approximately 54% of the aircraft's maximum fuel capacity. The ramp weight was 414,760 lb. (188,200 kg.). The 74% of maximum takeoff weight was chosen so that the aircraft would not be far above its maximum landing weight when we started to perform landings. Center of gravity at the blocks was at 23.5%.

The A340 cockpit arrangement and layout are almost identical to the A320, except for the obvious four throttles in the center console. The cockpit has a roomy feeling and uncluttered appearance, with the sidestick controller replacing the yoke found in many transport category aircraft. The electrical control for moving the left seat was especially helpful during the many pilot changes made during the flight.

Cockpit layout and instrumentation of the Airbus Industrie A340 are almost identical to those of the A320, except for the four engine throttles. The European consortium has had few requests from A320 operators to change the instrumentation of the aircraft.

The prestart checklist was accomplished quickly by Baud. The "lights out" philosophy of extinguishing all white lights on the overhead panel for normal operations has been carried over from other Airbus Industrie aircraft. Baud also inserted the data in the flight management and navigation systems that would keep us in the operational control area near Toulouse.

Both right engines were started simultaneously by Baud using the engine start switches behind the throttles. The engine start parameters were observed on the center engine/warning display monitor. I initiated the start of the No. 1 and No. 2 engines after the two right engines were stabilized. There was a short power interruption when the No. 1 generator came on line, but Baud said that would soon be corrected in the A340s.

The Garrett auxiliary power unit was used for the start and has ample power to start two engines at the same time. Baud said that during the 840 hr. of flight testing, Airbus has not experienced any problems with the APU. Additional power beyond idle was required to initiate the taxi to Runway 33L. Once moving, idle power was ample to provide a comfortable taxi speed for the relatively lightweight aircraft. The brakes were tested and deceleration was smooth without any tendency to grab. The more than 70 deg. of nosewheel steering provided through the steering handle was adequate to make the turns required during taxi.

A reduced power setting had been selected to give the same takeoff performance experienced during a maximum takeoff gross weight. The V_1 takeoff decision speed was calculated to be 126 kt. and the rotation speed was 128 kt. Takeoff safety speed (V_2) was 138 kt., and the three speeds were shown on the primary flight display. Total fuel used during the 6 min. from engine start to initiating takeoff was 600 lb. (270 kg.).

I advanced the four throttles to the flex detent for takeoff. Acceleration was smooth and rudder pedals were used for directional control throughout the takeoff roll. A total of 4,100 ft. (1,247 meters) was used to reach the rotation point at 135 kt. There was a tailwind of 9.2 kt. during the takeoff. Liftoff occurred at 5,110 ft. (1,555 meters) and at 142.6 kt. The pitch angle at liftoff was 10.3 deg. The 35-ft. altitude mark was reached 5,880 ft. (1,789 meters) down the runway at 147.4 kt. The pitch angle at that point was 11.5 deg. and the flight test data indicated an average 3.0 deg./sec. pitch rate. The fly-by-wire system has a maximum pitch rate limitation to eliminate the possibility of a tail strike at takeoff.

Once the landing gear and flaps were raised, a 250-kt. climb speed was established and the power levers were in the climb detent. An altitude of 10,000 ft. was reached in less than 6 min. after brake release and a total of 4,040 lb. (1,833 kg.) had been used since engine start. Rate of climb passing through 10,000 ft. was 2,800 fpm. The aircraft was leveled at 15,000 ft. to observe the high-speed protection afforded by the control laws applied to the A340. The A340 control laws are used to protect it in both Mach number and speed. The A320 only has V_{mo}, protection.

At 300 kt. the nose was depressed to a 5-deg. down attitude with power setting to achieve a 1 kt./sec. acceleration. When the speed indicated maximum speed plus 15 kt. and with full nose-down command on the side-stick controller, an automatic and progressive up elevator command was applied. The pilot nose-down authority is reduced at the same time to give a maximum stabilized speed of V_{mo} plus 15 kt. or M_{mo} of plus 0.04.

At the same altitude, the bank angle was increased to its maximum of 67 deg. while the power was automatically increased to maintain level flight. Once the side-stick controller was released, the aircraft's control laws automatically reduced bank angle to 33 deg. After bank angle is established below 33 deg., the angle will be maintained with the stick in the neutral position. As I had found in the A320, it is easier to fly with pressure on the top of the stick during maneuvers rather than grabbing the entire column.

The time spent at 15,000 ft. was about 5 min. A climb speed of 300 kt. was established to reach a cruising altitude of 35,000 ft. A little over 2.5 min. elapsed prior to reaching 20,000 ft. Rate of climb was 1,500 fpm and the total fuel flow was 27,800 lb./hr. (12,615 kg./hr.). Total fuel burn from blocks was 7,886 lb. (3,578 kg.).

It took 3.3 min. to reach 25,000 ft. from 20,000 ft.; the rate of climb was 1,600 fpm. It required another 4.2 min. to reach 30,000 ft., and the rate of climb was 1,000 fpm. The fuel flow was 21,123 lb./hr. (9,584 kg./hr.). Total fuel used from blocks was 10,910 lb. (4,950 kg.).

A stable cruise speed of Mach 0.83 at 35,000 ft. was established within 7 min. of passing through 30,000 ft. Total fuel used since en-

Standard interior for the A340-200 transport consists of 18 first-class seats, 42 business class and 203 economy-class seats. The longer A340-300 aircraft can accommodate nine more passengers in business class and 25 additional economy-class travelers.

gine start was 13,174 lb. (5,977 kg.). The amount of fuel consumed from brake release at takeoff to 35,000 ft. was 12,574 lb. (5,707 kg.), including the time spent at 15,000 ft. doing maneuvers.

The fuel flow at Mach 0.83 was 14,190 lb./hr. (6,438 kg./hr.). Aircraft weight was now 401,630 lb. (182,228 kg.). Baud said the nominal A340 cruise speed would be between Mach 0.82 and 0.84. The pitch angle of the aircraft at 0.83 was 2.1 deg. up.

At the Mach 0.84 cruise speed, fuel flow was 15,538 lb./hr. (7,050 kg./hr.). While Airbus officials would not discuss specific performance guarantees with its customers, they said the A340 would meet the guarantees based on test data. The guarantees are primarily based on range and payload requirements for individual air carriers.

The outside visibility from the left seat was excellent during the climb to altitude. I was able to observe the left wing tip and part of the No. 1 engine without resorting to contortions. The view over the nose also was very good. This permitted excellent situational awareness in the landing pattern and on landing later in the flight.

Airbus Industrie has carefully matched human engineering with situational awareness in its design of the instrument panel. I was able to quickly establish an efficient scan of the primary flight display to determine what the aircraft was doing at any one time. While I found the rate of climb indicator difficult to assimilate in my scan on similar displays on earlier flights, I was more comfortable with the display on this flight.

Another pilot occupied the left seat prior to descending to a lower altitude to evaluate the aircrafts slow-flight characteristics. The initial descent was made at 290 kt. with a 4,000-fpm rate. The speed brakes were deployed to the one-half position with a very slight buffet. The speed was 308 kt. and still with a descent. When the spoiler speed brakes were deployed to their maximum, the buffet increased, but was still mild. The rate of descent at 303 kt. was 5,400 fpm.

The A340's control laws provide high angle-of-attack protection, much the same as they did in the A320. The laws provide for the elimination of the risk of stall in high dynamic maneuvers or in gust or wind shear conditions. The design aim also is to provide the capability to reach and maintain a high coefficient of lift with the side stick in the full aft position without exceeding the stall angle.

During these low-speed flights flown by the guest pilot and Baud, the aircraft was put through a number of dynamic maneuvers with normal control laws applying and also with alternate and direct laws.

In normal laws, without the use of automatic thrust being applied, the control laws provide for angle-of-attack protection with the stick in the neutral position. A lower speed of approximately 10 kt. can be achieved with the stick in the full aft position.

The aircraft was completely controllable during these maneuvers up to the maximum 45-deg. bank angle allowed and 170 kt. and an angle of attack of 13-deg. During one dynamic maneuver performed by Baud involving an abrupt rolling pull-up, the aircraft experienced some buffeting. Baud said the maximum angle of attack limit would he lowered by 0.5 units so that even in an extreme dynamic maneuver there would be no buffeting during recovery. The load protection limitation was active during these maneuvers, holding the aircraft to a maximum of 2.5g. With slats down, the maximum is 2.0g.

A revision to the alternate control law requires two failures in the flight control system, and high AOA protection is replaced by the static stability of the aircraft. Baud said this places the aircraft in the flight characteristics mode comparable to transports without a fly-by-wire flight control system.

Further failures in the flight control system, which Airbus says is a 10-7 probability, would revert the A340 to the direct laws. Baud said this was equivalent to flying an aircraft without the stability augmentation system engaged. The stick is still operative but all control laws are lost. In order to reach the direct law status, Guyot had to artificially disable the Aerospatiale primary and Sextant/Honeywell secondary flight computers from the flight test station in the cabin. During a stall in the direct laws, the angle of attack reached 24.7 units with the flaps at 26-deg. The power was at idle and the stall speed was 107 kt. with a 2,000-fpm descent. The aircraft rolled slightly to the left at the stall before Baud initiated the recovery.

In the unlikely event that all four electrical generators were inoperative in flight, the APU generator can be used. If that were inoperative, the Sundstrand ram air turbine mounted under the wing would be available for electrical power.

The A340 was then flown to the Toulouse airport for landings on Runway 33L. The other pilots made a number of approaches to the runway with generally very smooth landings. The touchdowns on the Dowty articulated main landing gear were almost unnoticeable. The articulated landing gear allows the rear wheels to touch down first before the front wheels and oleo absorb the aircraft's weight.

First Airbus Industrie A330 is in final assembly at Toulouse, France, and first flight is scheduled for October. The twin-engine transport is almost identical to the four-engine A340-300.

Airbus Industrie A340 and A330 transports are assembled on the same production line at Toulouse. There is room for expansion, if future airline market demands dictate.

Touchdown on the center landing gear provided the same results. It was when the nose gear touched down that we realized we were on the runway. The nose gear landed hard on almost all the landings and otherwise spoiled a smooth landing. Baud said that minor modifications would be made to the Messier-Bugatti front gear to smooth the landings. The A340-200 has a negative 1.6-deg. pitch on the ground. The longer A340-300 does not have the same negative pitch and smoother landings are possible, he said. I took control of the aircraft after one of the touch-and-goes on the downwind leg. The speed control was set at 140 kt. and partial flaps were extended. At the fairly close abeam position for a large transport, I lowered the landing gear at the 180-deg. position and shortly after selected full flaps. The approach was flown at 140 kt., and I found that I had no tendency to overcontrol the aircraft. The A340 has the same stick control movement and forces as the A320, Baud said.

A slight flare was initiated at 40 ft. above the runway and the throttles were reduced to idle near 20 ft. The landing was smooth at 130 kt., and although I tried to lower the nose wheel slowly, there was still a relatively hard touchdown on the nose landing gear.

The ability to land the aircraft smoothly on the first attempt was naturally aided by the automatic speed control, allowing me to concentrate on positioning the aircraft vertically and in relation to the runway. The lack of throttle movement during the approach was not a factor in situational awareness. Flying the A340 with the side-stick controller on a visual approach gives the pilot a sense of immediate response and masks the fact that the A340 is a large four-engine aircraft.

During one of the other pilot's touch-and-go landings, Baud retarded the throttle on the No. 4 engine to idle at 600 ft. after takeoff. The sideslip (beta target) indicator below the roll indices on the primary flight display showed to the left and in a blue color. As the rudder was depressed to compensate for yaw, the target moved to the center and changed to a normal yellow color.

A two-engine out on the same side was not performed, but Baud said there was ample rudder authority to control the aircraft. At speeds below minimum control speeds, a slight bank angle was needed for heading control. At 155 kt., the aircraft could be flown with rudder alone, Baud said. Airbus officials were quick to point out, however, that their experience with the CFM International engines during flight tests on the A340 had been very positive, with no problems and excellent reliability.

Another of the guest pilots made the final landing and had no difficulty stopping the aircraft with the antiskid system and reverse thrust. Total air time was 2.2 hr., while the blocks time was 2.4 hr. Total fuel used during the flight was 34,735 lb. (15,760 kg.). Eight touch-and-go landings were accomplished during the flight.

Following the evaluation, Guyot said the flight test program had generally gone very well, with few aircraft or system modifications required. Early in the program, there was slight buffet experienced when accelerating beyond Mach 0.83. Wind tunnel testing identified airflow separation on the wing adjacent to the inside of the outboard engines. A blister, which is hardly noticeable from the ground, was added to correct the airflow. Another change was that the chord of the No. 1 slat was increased, resulting in a 1.5% drag reduction, Guyot said.

A320 operator survey

Airbus officials have surveyed operators of the A320 to determine whether modifications were required to the A340 to improve cockpit workload or operational performance.

More than 310 A320s have been delivered to some 30 operators since first deliveries in 1988. Baud said that at a meeting earlier this year with 27 A320 operators, there were no calls for substantive changes to be made in the A320. A few cited the need for a reduction in cockpit noise, easier autopilot and flight director selection, and the ability to detect if the pilot not in command is touching the side-stick controller. The A320 met the guarantees, so Airbus is looking at proposed fixes that would meet the suggestions, but at costs to the operators, Baud said. There were no demands for moving throttles in the automatic modes, he said.

Baud also emphasized that there were no changes mandated by the investigations following the three A320 crashes (AW&ST Mar. 23, p. 62). He stressed, however, that in new-technology aircraft, such as the A320 and A340, comprehensive initial and recurrent pilot training was an important factor in safe operations.

Airbus holds 115 orders for the A340—about one-third for the shorter, long-range A340-200 and the remainder for the A340-300. The approximate price for the A340-200 is $105 million, and $110 million for the A340-300.

Route evaluations are planned to begin in November. Air France Group will operate an A340-300 from its Paris hub and Lufthansa an A340-200 from Frankfurt. Lufthansa is scheduled to receive its first A340-200 in January, 1993, and shortly after Air France Group will take delivery of an A340-300.

With the introduction of the A340 and later the A330, Airbus Industrie will be able to offer operators a wide range of aircraft from the A340-200—with its 262 seats and 7,500-naut.-mi. range—down to the smaller, shorter-range A320 series.

AIRBUS INDUSTRIE A34-200 SPECIFICATIONS

POWERPLANTS

Four CFM56-C turbofan engines with a 31,200-lb. thrust rating. An option of an uprated engine to 32,500 lb. is planned by the end of 1993.

DIMENSIONS

Length	195.0 ft. (59.4 meters)
Fuselage dia.	18.5 ft. (5.63 meters)
Overall height	55.2 ft. (16.8 meters)
Wingspan	197.8 ft. (60.3 meters)
Wing area	3,910 sq. ft. (363 sq. meters)
Wing sweep	30 deg.

Note: The length of the A340-300 is 20.9.0 ft. (63.7 meters)

WEIGHTS

Maximum takeoff weight	558,900 lb. (253.5 tonnes)
Maximum landing weight	399,000 lb. (181.0 tonnes)
Typical operating weight empty	270,700 lb. (122.8 tonnes)
Maximum payload	101,800 lb. (46.2 tonnes)
Maximum fuel capacity	238,900 lb./35,660 gal.

Notes: The landing weight, operating weight and payload of the A340-300 is higher than the A340-200.

Higher maximum takeoff gross weight versions of both the A340-200 and A340-300 are expected to be available in 1995/1996.

PERFORMANCE

Takeoff field requirement, sea level ISA plus 15C, 6,000 naut. mi. range	7,600 ft.

Maximum range with 262 passengers, baggage	7,500 naut. mi.
Maximum Mach number	0.85
Typical passenger loading, three classes	262
Underfloor capacity	27 LD3s or nine pallets

Extensive MD-11 automation assists pilots, cuts workload

David Hughes/Yuma, Ariz.
October 22, 1990

On the McDonnell Douglas MD-11, computers perform everything from checklist tasks to stall recovery in an aircraft where the flight engineer's duties are accomplished by a bank of automatic controllers that run the aircraft's systems.

The cockpit design distills the experience of 19 years and more than 16 million hours of DC-10 commercial airline operation into computerized system controllers that operate hydraulic, electrical, air (pneumatic) and fuel systems. These aircraft system controllers, as they are called, run in parallel rather than in series for both normal and emergency procedures. This means, for example, that the fuel system configuration changes to the proper status without waiting for steps to be taken to reconfigure the hydraulic, electrical or air systems. Each system is run by two computers and each one can revert to manual operation if necessary.

In addition, the automatic flight system (AFS) includes augmentation in pitch, yaw and roll. Roll control wheel steering is optional. The autothrottle portion of AFS keeps the pilot from unintentionally flying too fast or too slow in a particular configuration. A flight management system provides automatic navigation in both vertical and lateral dimensions. It also provides the optimum speeds and altitudes to achieve the most efficient fuel consumption on a particular route. The two dual-channel flight control computers direct the throttles, the ground spoilers, the elevators, ailerons, rudder, elevator feel, flap limiter and stab trim.

This *Aviation Week & Space Technology* pilot evaluated the MD-11 recently from the left seat in test aircraft No. 4. The aircraft, which is powered by General Electric CF6-80C2 engines, is scheduled for delivery to American Airlines later this year.

Capt. John Miller, chief of flight operations for the MD-11 program, occupied the right seat for the demonstration. The flight, which was part of the regular test program, lasted 5.8 hr. and more than 40 test procedure cards were accomplished. A crew of technicians then manned computer and video consoles in the back of the aircraft to monitor the tests in progress, which included numerous checks on the accuracy of the Honeywell flight management system.

The exterior of the MD-11 looks very much like a DC-10 because the airframe is stretched only 18 ft. 6 in. The MD-11, however, is most recognizable by its winglets. The outer portion of the wings have a blunt trailing edge that varies in thickness from .5 in. near the wingtip to 1.5 in. at the inboard edge of the outer flap panel. The inlet for the No. 2 engine is the same as that used on the DC-10 Series 40. This is the case even though the MD-11's 61,500-lb.-thrust General Electric CF6-80C2 and 60,000-lb.-thrust Pratt & Whitney PW4460 engines require more airflow than the 52,500-lb.-thrust Pratt & Whitney JT9D59A series engines on the DC-10-40. McDonnell Douglas engineers were concerned about whether the opening would be large enough, but flight tests have demonstrated that the air mass movement is sufficient. The third engine in the program is the 65,000-lb.-thrust Rolls-Royce Trent 665. Composite material is now used in the No. 2 inlet structure.

Once we were in the seat, Miller turned on the inertial systems and called up route F-150 in the flight management system (FMS) memory. F-150 is a round-robin route that goes from Yuma to San Diego, Los Angeles and Palmdale in California and then on to Boulder, Colo. This route would be used to test the FMS navigation accuracy. Miller entered a zero cost index to govern FMS calculations. An index can be selected between 0 and 999, depending on whether fuel economy is the primary objective or whether reducing time en route is more important.

The next entry in the FMS control display unit was our gross weight of 460,900 lb. and total fuel of 186,400 lb. with a zero fuel weight center of gravity of 23%. At this point the FMS calculated our c.g. with full fuel.

Miller filed to climb to a cruising altitude of flight level 260 and the FMS computer told us that flight level 340 was optimum for this fuel weight and flight level 366 was the maximum level. One of the features of the MD-11 FMS that international operators will find useful is the system's ability to calculate a series of step climbs with up to six steps. This allows for more accurate fuel planning on long-range flights.

The aircraft system controllers are programmed to perform self-tests. The fuel system, for example, completes its test when the refueling door is closed. Miller checked with the scanner on the ground that the flight controls were clear before initiating an automated hydraulic system test that would move control surfaces. If any problems are discovered they are annunciated to the crew.

A check of the fuel system synoptic page, a diagram with tanks and pumps and engines depicted, showed that the No. 1 main tank contained 41,100 lb., the No. 2 main tank 64,600 lb., the No. 3 main 40,900 lb. and the auxiliary tank 39,800 lb. The No. 2 tank is really two tanks in the inboard sections of the left and right wings, and the auxiliary tank is in the center wing. This diagram is a dynamic one that tracks fuel levels and notes with changing colors when fuel pumps are on or off. The fuel quantities are measured by a computerized system that relies on several probes in each tank. The tail tank, which has a 2,000-gal. capacity, was empty but the fuel system controller would route fuel into it after takeoff to achieve the desired aft center of gravity.

A check of the configuration page on the Systems Display cathode ray tube (CRT) showed that the brakes were at about 27C and the tire pressure was about 190 psi. A check of the overall system status page showed that there were no alerts with consequences.

Miller depressed the APU start button, and the APU doors opened and the power unit began to turn over. He then called up the fuel status page to show that a fuel pump had been turned on automatically to feed the APU and the fuel system configuration was changing. He called up the electrical system status page next and showed that the APU was automatically powering the main buses and the electrical system was reconfiguring itself to draw power from the APU. External power was turned off and the air cycle machines began operating automatically.

Manual-automatic shifts

Miller demonstrated that it is possible to intervene at any point and convert the fuel, electrical, air or hydraulic system to manual operation if the pilot so desires. Reverting to automatic operation is as simple as depressing a button.

The crew entrance door was closed and the FMS calculated all of our takeoff speeds, which appeared as markers on the speed tape. V_1 (takeoff decision speed) was 134 kt., V_r (rotation speed) was 150 kt. and V_2 (takeoff safety speed) was 163 kt. These markers would begin moving down from the top of the display as we accelerated and approached the speeds involved.

Miller planned to have me fly the takeoff with the autopilot off and the autothrottles engaged. We planned to take off on Runway 3L, which is 13,299 ft. long, and fly on runway heading before turning left to Bard, which is the VOR located near Yuma.

Miller pushed the No. 3 ignition switch, and the air system reconfigured itself by shutting off the packs to provide bleed air from the APU for engine start. He then pulled out the start switch. N2 started to accelerate and when it reached a blue line on the engine and alerting display at about 15%, Miller pulled out the electrical fuel switch to turn it on. A line, which represents the starting limit, appeared on the exhaust gas temperature gauge on the engine and alert CRT. This line goes away after engine start is complete. Light-off occurred and the engine accelerated to about 45% of N2. The air, hydraulic, fuel and electrical systems were automatically reconfiguring as the start sequence progressed.

After starting engines 1 and 2, Miller said, "The electrical system has picked up on the buses and the tie bus. The air system is now giving us air conditioning. The fuel system is transferring fuel." I then started the No. 2 engine, and Miller shut down the APU.

Miller pointed out one feature of the fuel and hydraulic synoptic pages that serves as a good reminder to the pilots later in the flight. At the time of engine start, a blue line appears at the top of the three hydraulic system reservoir diagrams to show where the quantity is at engine start. Should fluid be lost during the flight, the discrepancy is noted by the difference between the fixed blue line and declining fluid level marked in solid gray. The same sort of blue line also appears on the oil quantity diagrams. Miller used an abbreviated checklist that fits on one side of a laminated card to double-check critical items.

I released the brake and began taxiing the aircraft. The nosewheel steering control on the left side was easy to use; however, I had to work on my taxi speed and rate of turn to come around a corner smoothly because I was unfamiliar with the aircraft. Nosewheel steering provides 67 deg. of authority left and right, and rudder pedal steering provides 12 deg. of authority left and right. One feature on the primary flight display that was helpful was a readout of taxi speed. I kept the MD-11 moving between 10 and 20 kt.

Clearance to Bard

We were cleared into position on Runway 3L with the winds at 060 deg. and 8 kt. The takeoff would be made with bleeds off and air cycle machines off. We were cleared to fly runway heading to a point 3 naut. mi. past the airfield boundary before turning left to Bard. Our assigned altitude was 4,000 ft. and we were told by ATC to expect clearance to flight level 260 10 min. after departure. We used a call sign of DACO 450, which was derived from Douglas Aircraft Co. aircraft number 450.

We were cleared for takeoff. The c.g. was 23.1% and the flap setting was 15 deg. for the derated thrust takeoff. The takeoff run began at 11:15 a.m. as I began advancing the throttles. With autothrottles on, the automatic system takes control of the throttles when the engines reach 60% of N1 and sets takeoff thrust.

McDonnell Douglas modified the DC-10 Series 30 wing design for the MD-11. Leading edge slats extended automatically if the aircraft approaches a stall. Another automatic feature involves flap retraction to avoid an overspeed condition.

The takeoff thrust target is marked with a "V," which rests on the outer edge of the three round-dial N1 diagrams on the engine and alert CRT. A line that ends in a T moves up as power is advanced, and when the desired setting is reached the T fits inside the V to provide the pilot with a simple visual cue to show that the power setting is correct. The pilot can always override the autothrottles and push the power up to maximum rated thrust at the forward physical stop. The throttles can be pushed past this gate in an emergency with a 30-lb. force to achieve the maximum available thrust limited to engine red line.

The aircraft reached 80 kt. 7 sec. after the final thrust setting was made and the autothrottles entered CLAMP mode, which meant the throttles were fixed for takeoff and could not roll back. Nine seconds later we reached the V_1 speed of 134 kt. at which point we were committed to the takeoff. Four seconds later I pulled back on the yoke and the nose rotated smoothly into the air as I began following the rising pitch V bar. We had used about 4,000 ft. of runway on the takeoff roll. Above the flight director V bar was a pitch limit indicator that marked the not-to-exceed angle of attack. This symbol would change from blue to amber if we approached within 2 deg. of the stick shaker angle of attack, and it would turn red if we were about to reach the stick shaker angle itself.

A pitch limiter is a standard feature for wind shear escape maneuvers, but the MD-11 can display it during normal operations as well. Miller noted the positive rate of climb and put the gear handle up. As each V speed was reached, the associated marker would come down the airspeed scale.

The aircraft climbed rapidly as it had plenty of excess power considering it was nearly 160,000 lb. below the 618,000-lb. maximum allowable takeoff gross weight. I rotated the nose up to about 25 deg. to follow the V bar, which was calling for the maximum pitch the flight guidance system ever commands. The aircraft continued to accelerate to V_2 and then to 173 kt., or V_2 + 10.

At 400 ft. above the ground, Miller engaged the autoflight system and the profile VNAV mode and the autopilot continued the climbout as I monitored the controls. At 1,500 ft. above the ground, the autothrottle system reduced the power to a climbout setting without any further action from me. At 3,000 ft. above the ground, the aircraft nosed over and Miller retracted the flaps as we began to accelerate from 173 kt. to 250 kt. for the climb profile to 10,000 ft. The MD-11 will fly this portion of the profile at 1.3 V_{stall} +5 kt. if that happens to be higher than 250 kt. for a given weight and set of conditions.

At 213 kt. (V_2 + 50 kt.), Miller retracted the slats. With the lateral NAV mode engaged, the automatic system had already initiated a left turn to proceed to Bard VOR and we were cleared up to 7,000 ft. Setting a new altitude limit into the FMS is a simple procedure using the controls on the glare shield.

The flight control panel on the glare shield allows the pilot to make changes in heading, airspeed and altitude and to alter these values in the flight management system computer without going through the two multifunction control display units on the pedestal. The pilot simply changes the preselect value in a window and pulls a knob out to activate a new heading, airspeed or altitude. To hold the current speed, heading or altitude, the pilot simply pushes in on the same knob. FMS push buttons under the altitude, speed and heading knobs allow the pilot to turn the knobs and then send these changes to the FMS for use in its calculations. If the pilot depresses one of the FMS push buttons without turning the knob, the value in the FMS program will be retrieved for use.

"You don't have to get down to the control display unit and fiddle around because there is a great deal of interface between the glare shield and the FMS," Miller said. The glare shield system directs the changes to take place regardless of the FMS or automatic flight mode engaged.

Five minutes after takeoff, Miller called up the fuel system synoptic page on the systems CRT and I noticed that the fuel system had automatically pumped 3,900 lb. of fuel into the tail tank and that our c.g. had moved aft to 25.6%. The MD-11 is designed to be flown at an aft center of gravity in cruise to reduce drag. After passing 10,000 ft., the system established the best economy climb speed of 347 kt.

After passing Imperial VOR as we were flying west to Kumba intersection, we were told by ATC to maintain our present heading. This was accomplished by pushing the heading knob on the glare shield to maintain the heading, and the deviation to the left of our intended course was depicted clearly on the navigation display.

Later, as we returned to course and passed over Julian, the autopilot smoothly banked 10 deg. to turn to a heading of 294 deg. The navigation display provided a wealth of data. All six CRTs were easily readable in the strong sunlight, and the color legends made it easy to interpret data.

All speeds, headings and altitudes specified by the pilot are displayed in white, for example, while all speeds, headings and altitudes derived from the FMS are shown in magenta. This helps the pilot remember when he has intervened in the automatic operation and when the FMS is navigating. It is possible to call up a standby flight plan on the navigation display in a second color for planning purposes.

The FMS automatically tunes the appropriate navaids for the route being flown. The system relies on three Honeywell ring laser gyros as well as two scanning DMEs with five channels each—one for VOR, one for ILS and three for the FMS to use in precision navigation. The pilot can take control of the VOR/DME or the ILS/DME if he prefers. The MD11 has two VORs, two ILSs, two automatic direction finders and provisions for two microwave landing system receivers as well. Global Positioning System capability can be added in the future by inserting a circuit card in the inertial reference unit.

We passed over Catalina Island at flight level 260 and Mach 0.808 and were cleared up to flight level 270 as we turned north toward Los Angeles. Miller entered the new altitude assignment into the FMS from the flight control panel on the glare shield. The AFS vertical alert warned us that the system was initiating a climb to flight level 270, so we could override it if we wanted. We burned an average of 18,000 lb. of fuel an hour during the en route cruise portions of our flight.

Performance improvements

The MD-11 has a range shortfall owing to lower than expected fuel efficiency on the General Electric and Pratt & Whitney engines and higher than expected aircraft weight. The engine manufacturers are working to correct the 4-5% specific fuel consumption shortfall, while Douglas is working to decrease the aircraft's empty weight (AW&ST Aug. 6, 1990, p. 70). Douglas may also achieve some small improvements by fine-tuning the aerodynamics, according to Miller.

The empty weight has been reduced by 1,700 lb. so far, and the maximum allowable takeoff gross weight being offered as an option has been increased as part of the effort by Douglas to meet various payload guarantees over defined routes. Douglas just added another 3,000 lb. to bring this optional weight to 618,000 lb. At this weight, the aircraft will be able to fly nearly 8,000 stat. mi., and some airlines are expected to opt for the higher weight. The standard maximum takeoff weight remains 602,500 lb.

At this point, other Douglas test pilots on the flight took turns in the left seat to perform additional tests on the FMS during the round robin back to Yuma. The MD-11 completed the circuit to Yuma and headed back to California and out over Mission Bay to Warning Area Whiskey 291 over the Pacific Ocean.

MD-11 fuel system controller transfers fuel to the tail tank during cruise to achieve an aft center of gravity for the best fuel economy. The fuel is pumped forward prior to landing.

After getting back into the left seat, I glanced at one of the CRTs to update myself quickly on the status of the flight. The progress page, as it is called, consolidates on one CRT display all of the data needed for an International Civil Aviation Organization (ICAO) position report, plus a lot of other useful information. In addition to the last position, time and altitude, it gives the next position, estimated time of arrival and altitude and the position after that. The page also records the outside air temperature, wind, fuel remaining, distance to go to destination and the fuel that will be remaining at the time of arrival. It also gives a distance to the alternate airport, estimated arrival time and fuel that will be remaining there. "This is an example of how we have designed the system to serve the pilot," Miller said. In earlier-generation aircraft, pilots had to hunt for this type of data before making a position report.

I noted our fuel on board was 113,300 lb. and we would have 99,100 lb. left on arrival at Yuma, 322 naut. mi. away at 4:05 p.m. To keep track of our position within the warning area, Miller called up a flight plan loaded with the points on the area's boundary. When this flight plan was displayed on the navigation display unit, the connect-the-dot pattern depicted the Whiskey area and how close we were flying to the area's borders.

I disconnected the autopilot and an autopilot-off message surrounded by a box that was flashing appeared on the primary flight display. A touch of a button acknowledged that I was aware the autopilot was off and the flashing stopped. I began to fly the aircraft at 17,500 ft., and when the aircraft was 150 ft. below the designated altitude in the FMS a chime sounded and a computer voice said "altitude." The MD-11 offers a variety of voice warnings that customers can select, including 1,000 ft. above a level-off. If the aircraft is climbing or descending too rapidly as it approaches a level-off, a voice warning will be activated.

The autothrottles were holding the speed at 220 kt. I disconnected that system and an autothrottles-off message appeared in a flashing box. A speed bug on the speed tape provided an easy peripheral visual reference as to whether I was flying the desired speed ±5-10 kt.

As the aircraft decelerated slightly, a green bar extending down appeared, predicting the speed I would be flying in 10 sec. I added a little power to maintain speed, and Miller turned off the longitudinal stability augmentation system (LSAS), which uses ±5 deg. of elevator deflection to augment longitudinal control. With LSAS on, the pilot sets a pitch attitude and the LSAS system holds it with elevator inputs. When LSAS is turned off, the aircraft still flies smoothly, but I needed to trim the aircraft to maintain the selected attitude. "It's just a regular airplane now," Miller said. I experienced no difficulty flying the aircraft at an aft center of gravity and a slow speed.

Miller reengaged the LSAS to demonstrate its speed protection features as I began to slow the aircraft in a figuration toward stick shaker speed. Miller suggested that I attempt to close the throttles and maintain a 10-deg. pitch attitude. The first protection feature to be activated was the autothrottle system, which started to advance the throttles to maintain V_{min}. This process would have continued until the autothrottles reached maximum continuous thrust. To override this, I held the throttles back against pressure.

The LSAS then activated its V_{min} protection and began pushing the yoke forward. "There are two protections for every condition," Miller said. To slow enough to reach the stick shaker speed of 150-160 kt. in a clean configuration, I had to hold the yoke back against substantial pressure as well.

Five MD-11s are involved in the flight test program at Douglas' Yuma facility. Ship 4 (foreground) is dedicated to avionics testing; Ship 3 (background), to airframe/engine testing.

The pitch limit indicator tuned amber and Miller estimated the LSAS was providing 50 lb. of forward pressure on the yoke by this time. The PLI turned red as the stick shaker began to vibrate the yoke.

By releasing the yoke and throttles, I allowed the aircraft to recover on its own. The LSAS system lowered the nose and the roll control wheel steering kept our bank angle under 5 deg. during the recovery as the autothrottle system selected maximum continuous thrust. The slats also extended automatically as noted by a legend on one CRT, and we felt a little bit of shuddering. Our altitude loss was negligible. The slats retracted when a safe speed was achieved. Miller later explained that if the maximum speed allowable is about to be exceeded, the protection systems throttle back and raise the aircraft's nose.

Miller then demonstrated how the speed tape keeps the pilot informed of his changing minimum and maximum speeds, depending on the aircraft's configuration. Two amber-colored areas marked on the top and bottom of the tape move whenever the configuration changes to show where the speed limits are located. The lower amber region, for example, starts at 1.3 V_s for the current configuration and extends down to the speed marked in red where the stick shaker would activate. The speed command bug cannot be moved any closer than 5 kt. above the minimum or 5 kt. below the maximum speed, even if the pilot inadvertently tries to select an inappropriate speed.

Miller showed how the top of the slow-speed amber region dropped from 197 kt. to 160 kt. as he extended the slats and lowered the aircraft's 1.3 V_s speed. Another marker showed the speed at which we could safely retract the slats. As I slowed the aircraft, Miller extended the flaps to 28 deg. and the top of the slow-speed amber region moved down again to 153 kt. He then extended the flaps to 35 deg., and I slowed the aircraft to 120 kt. When the stick began to shake, I released the controls and the aircraft automatically recovered.

Following this, Miller extended full flaps to 50 deg. and the maximum speed for leaving the flaps out was marked on the tape at 175 kt. As I accelerated toward this speed, the flaps began to retract automatically to 40 deg. When Miller raised the flaps to 28 deg., the maximum speed marker for leaving the flaps out moved up to 210 kt.

Following this slow-flight sequence, we engaged the flight management system to take us toward San Diego climbing back to flight level 270 at a speed of 325 kt., the economy speed for our weight. As we headed east back to Imperial, we encountered a line of thunderstorms and the weather radar painted heavy rain shown as red on the navigation display CRT. The storms were ahead and to our left. Miller later explained that the intensity of the weather presented on the navigation CRT can be adjusted independently from the intensity of the navigation map display even though the two appear together.

Weather can be overlaid on an HSI or any other navigation display. As the tops were above our altitude, we deviated to the southeast. We were approaching the top of descent as calculated by the FMS.

We began a descent to flight level 240 until ATC held us at flight level 250 for traffic. Miller said the top of descent points used in test flights had worked out well, but in this situation ATC was keeping us higher than we wanted to be. As we proceeded on at flight level 250, an altitude message appeared to indicate we would not be able to reach our desired altitude of 2,500 ft. at Bard VOR, our initial approach fix. Miller began slowing the aircraft below 300 kt. and then a speed error message appeared indicating we would not be able to slow to our speed target of 180 kt. by Bard.

Nearing Imperial we were permitted to descend and Miller deployed the speed brakes, consisting of 30 deg. of spoiler extension, and picked up a 2,000-fpm rate of descent. He noted that it is possible to use both slats and spoilers at the same time. The aircraft could descend at 4,500 fpm at about 335 kt., but we were nearing 10,000 ft. at this point and we would have to slow to 250 kt. anyway. We slowed to 209 kt. at 11,000 ft. and Miller lowered the landing gear to increase drag.

The navigation display of our approach into Yuma showed a right turn to the downwind, a left base and the final approach leg. A blue ar-

Douglas plans to deliver five MD-11s to airlines by the end of this year. Two of the aircraft will be delivered to Delta Air Lines, two to Korean Air and one to Finnair. Aircraft production rates are expected to reach 1.25 per week or 62 per year by early 1992.

row along that route of flight showed where we would reach 5,000 ft., our next altitude limit. The display also showed distance and time to waypoints on the approach. We did not reach 2,500 ft. by Bard but we did descend to 1,700 ft. on downwind, so we managed to get down in time without additional off-course maneuvering.

As we passed the outer marker inbound on the approach, I disconnected the autopilot and began flying an ILS approach to Runway 3L with a V_{min} +5 kt. approach speed of 140 kt. With LSAS engaged, it was easy to hand fly the approach. Weather was not a factor because the sky was clear and the winds were relatively calm.

I went a little high on glideslope but corrected back. At 200 ft. above the runway, a computer voice began counting off the altitude based on a combination of radar and barometric altimeter inputs. The voice started at 200 ft. then read out 100 ft., 50 ft., 40 ft., 30 ft., 20 ft. and 10 ft. It took about 20 sec. to go from 200 ft. to touchdown. Miller said Douglas pilots can tell from the pacing of the readouts whether the touchdown will be a good one.

The touchdown was smooth but I was slow in lowering the nose gear to the runway. I depressed the go-around button on the throttles as Miller reset the flaps, and then I advanced the throttles. We rotated and climbed about 18 sec. after touchdown, with autothrottles engaged. Miller raised the gear and had to remind me to follow the single-cue flight director, which was providing takeoff guidance. He engaged the autopilot before we reached our level-off altitude. We planned a full stop and a taxi back for another takeoff on our next approach.

MCDONNELL DOUGLAS MD-11 TECHNICAL DESCRIPTION

CAPACITY

Passengers	250-405 (up to 405)
Lower deck (cargo)	6,850 cu. ft. (194 cu. meters)

DIMENSIONS

Wingspan	169 ft. 6 in. (51.7 meters)
Length overall	200 ft. 10 in. (61.2 meters)
Height overall	57 ft. 9 in. (17.6 meters)

WING AREA

Including aileron	3,648 sq. ft. (339 sq. meters)
Sweepback	35 degrees

LANDING GEAR

Tread (main wheels)	34 ft. 8 in. (10.6 meters)
Wheel base (fore & aft)	80 ft. 9 in. (24.6 meters)

ENGINES

Pratt & Whitney, PW4460	60,000 lb. max thrust
General Electric, CF6-80C2	61,500 lb. max thrust
Rolls-Royce, Trent 665	65,000 lb. max thrust

STANDARD DELIVERY WEIGHTS

Design gross maximum takeoff weight	618,000 lb. (280,321 kg.)*
Maximum zero fuel weight	400,000 lb. (181,437 kg.)
Operating empty weight	288,880 lb. (131,036 kg.)
Fuel capacity	40,183 U.S. gal. (152,109 li.)
Fuel weight	269,226 lb. (122,121 kg.)

PERFORMANCE

Weight limit payload	112,564 lb. (51,058 kg.)
Maximum level flight speed (31,000 ft.)	Mach 0.87 (588 mph) (945 km./hr.)
FAA takeoff field length (MTOW, sea level, temp. 30C)	10,500 ft. (3,200 meters)
FAA landing field length (MLW,Sea Level)	6,450 ft. (1,966 meters)
Design range (with FAR international fuel reserves):	
323 passengers and bags, 2 class	7,810 stat. mi. (12,566 km.)
293 passengers and bags, 3 class	7,980 stat. mi. (12,840 km.)

L-1011 cockpit automation cuts crew workload

Robert R. Ropelewski/Palmdale, Calif.
June 18, 1979

Commercial air transport operations are approaching a new level of cockpit automation with the entry into service of the first British Airways Lockheed L-1011-500 this month. Though the initial L-1011-500s delivered to British Airways are not equipped with the full line of automatic systems being developed for the aircraft, subsequent aircraft for the British carrier and for other TriStar 500 operators will incorporate systems aimed at reducing flight crew workloads and operating costs of the L-1011.

Primary new features incorporated in the L-1011-500 to attain these objectives are a flight management computer that combines both horizontal and vertical navigation modes, active aileron controls that reduce structural weight requirements while improving ride qualities and an automatic takeoff thrust control system permitting derated-thrust takeoffs, thereby extending engine life and reducing engine maintenance costs.

This *Aviation Week & Space Technology* pilot recently flew Lockheed's L-1011 development aircraft, the No. 1 TriStar prototype, in a 3-hr. test flight to observe the capabilities of these systems. Lockheed engineering test pilot Ebb Harris was in the left seat.

The flight management system was of particular interest, in view of today's fuel shortages, because of the possibilities it offers for additional fuel savings in the climb, cruise and descent phases of commercial transport flights.

Box-shaped route

The *Aviation Week* demonstration flight was conducted from Lockheed's production and flight test facility here and followed a box-shaped route from Palmdale to Santa Barbara, Calif., southwestward over the Pacific Ocean, then southeastward to Los Angeles International Airport and finally back to Palmdale, where several automatic approaches and landings were performed.

The TriStar developmental aircraft is equipped with a digital active control system for the ailerons, and wingtips that have been extended 4.5 ft. on each side from the standard L-1011. The cockpit was outfitted with the flight management system developed jointly by Lockheed and the Arma Div. of Ambac, Inc. (AW&ST Nov. 7, 1977,

p. 131), along with the standard TriStar avionic flight control system area navigation and automatic landing system.

The flight management system consolidates all these and adds the additional capability of computing and monitoring the optimum power settings for climb, cruise and descent. In each of these phases, the system offers at least two options the pilot may select—either minimum fuel or minimum-cost speed and thrust. The minimum cost option takes account of en route fuel costs and crew costs, among other factors, and generally offers a faster speed than the minimum fuel option.

The computer used with the flight management system is an expanded version of the area navigation computer used in the L-1011. Memory has been increased from 12,000 words in the navigation computer to 36,000 words in the flight management computer. In aircraft with the flight management system, the same computer serves both that system and the navigation system.

Other system elements

Other elements of the system are a central air data system, navigation receivers and aircraft sensor information from the engines. Each pilot has his own control and display unit (CDU) to communicate with the computer. The CDUs are located at the forward position on the center console.

Flight information is entered in the computer through the CDUs prior to taxiing for takeoff. In normal airline service, much of the information, like route lengths, refueling stops and fuel costs, can be stored permanently in the computer. The system in prototype No. 1 was programmed around Saudi Arabian Airlines' route network. Saudia was the first airline to have the flight management system installed in its L-1011s.

General aircraft performance data also are stored permanently in the computer memory. With the preprogramed information already in the computer, the pilot then adds such information as aircraft gross weight, fuel carried, proposed altitude with reported temperature for that altitude and desired Mach number.

Having received the constants, the computer, through the small CRT displays, then methodically explores options with the pilots, moving from one set of questions to the next as the pilot responds to the questions. These include selections between minimum-cost climb or derated thrust climb, for example, and between minimum fuel or minimum cost cruise.

The end-of-descent point—most likely the initial approach fix for the destination airport—is also entered, along with the desired airspeed and altitude at that point. During the cruise phase, the system advises the pilot of the optimum point to begin his descent to arrive with minimal fuel consumption at the end-of-descent point selected.

The computer serves an advisory function at almost every step in the programing. In the pretaxi programing in this case, for example, the computer challenged the selection of 35,000 ft. as a cruise altitude, recommending 37,500 ft. instead as the optimum minimum fuel altitude.

This capability to compute the most efficient altitude for a given aircraft weight at a given Mach number or speed schedule is one of numerous possibilities the system offers. Other information the system is capable of determining from the given parameters includes:

- Optimum Mach number for the assigned flight level. It can provide the datum input for this to the speed stability (Mach hold) system.
- Time lost by flying optimum Mach number vs. any other Mach number.
- Mach number optimized as a tradeoff between the fuel saving and schedule time loss.
- Excess fuel used by not flying at optimum altitude.
- When an altitude change is desirable, it can advise the crew to obtain an air traffic control clearance for the change.
- Favorable wind required at various altitudes to obtain equivalent fuel efficiency.

This information is calculated continuously by the computer, but is displayed on the CDU or used to control the aircraft only on pilot request.

Engine wear reduced

Takeoff was made from Palmdale using the L-1011's automatic takeoff thrust control system (ATTCS). On occasions such as this when the aircraft's takeoff weight is less than maximum, the system calculates a reduced takeoff thrust setting (engine pressure ratio, or EPR), imposing less wear on the engines. The derated thrust takes into account the possibility of an engine failure, and an engine failure after V_1 (maximum speed at which takeoff can safely be aborted) does not require any throttle increase by the pilot on the remaining two engines.

Shortly after takeoff, climb mode was engaged on the automatic flight controls and the aircraft stabilized in a 250-kt. climb. At 10,000 ft., the nose dropped slightly to reduce climb rate to 500 ft./min. while airspeed built up to the normal climb schedule speed at that altitude of 300 kt. The climb then resumed with the system holding both the speed and engine pressure ratio climb schedules.

The automatic flight management system offers several alternative climb modes:

- Minimum cost, which is a compromise between fuel costs and other operating costs.
- Minimum fuel, which is a computation of the minimum fuel flow rate that still provides an adequate climb gradient.
- Fixed climb rate of 250-300 kt./Mach 0.80.
- Manual throttle setting using derated thrust levels.

Approaching 20,000 ft. in our climb, the climb schedule was interrupted and altitude hold was selected for that level to demonstrate that interim altitudes can be captured and held by using normal autopilot altitude select procedures. The aircraft transitioned smoothly to level flight while maintaining the climb speed. The automatic climb mode was then reselected and a climb was reentered to 35,000 ft.

VOR stations selected

VOR navigation mode of the automatic flight control system (AFCS) kept the aircraft on a northwesterly track toward Santa Barbara, automatically seeking the best VOR frequencies for navigational fixes. The system seeks VOR stations as close as possible to 90 deg. from the heading of the aircraft for highest accuracy in position fixes, but it will accept any station from 20 to 60 deg. off the nose if necessary. Above 18,000 ft., the system will not use low-power VOR stations for position fixes.

The aircraft crossed the airspeed/Mach transition line at approximately 33,000 ft., and the climb increased significantly for a brief period as the system held Mach 0.80. Shortly afterward, however, the aircraft approached 35,000 ft. and the nose dropped smoothly to level the L-1011 at that altitude.

Airspeed built to Mach 0.85 and the throttles moved back slightly to hold the aircraft at its computed minimum-cost cruise speed, which is based on gross weight, selected cruise altitude, actual wind components, fuel consumption rate, fuel costs at the refueling site, hourly direct operating costs and stored aircraft/engine performance data. A headwind of approximately 63 kt. was indicated by the TriStar's inertial sensing system.

Outbound over the Pacific from Santa Barbara, a lower cruise altitude of 31,000 ft. was selected. Throttles moved back and the nose dropped slightly to establish a rate of descent of approximately 3,800 ft./min. while maintaining a speed of Mach 0.85. The system held the speed within ±0.005 Mach during the transition from cruise to descent to cruise again.

As the aircraft leveled itself at 31,000 ft., throttles came forward to establish the newly computed minimum-cost cruise speed for that altitude. Inertial sensing units indicated a headwind of 77 kt. at that altitude.

VOR reception was becoming intermittent on most stations at this point because of our distance from land. Santa Barbara VOR signals remained strong, however, and the flight management system went to a single fix on the Santa Barbara station.

As the aircraft crossed the westernmost waypoint on our route, it banked southeastward to pick up the next leg. As it did so, our former cruise altitude of 35,000 ft. was reselected on the AFCS and a climb was initiated to return to that altitude. Shortly after leveling off again at 35,000 ft., the aircraft passed through a wind shear in which the strong westerly winds dropped to about 30 kt. A minimum-cost cruise profile was still being followed, and the airspeed indicator read Mach 0.835.

A critical aspect of the flight management system, and a key argument for its use, is the precise control of aircraft speed it provides in the climb, cruise and descent phases of flight. It does this through constant monitoring and fine trimming of engine thrust.

Lockheed engineers concede that flight crews could do this manually, but they argue it would require continuous attention to the task of throttle management on the part of the crews. This is because power requires changes constantly as fuel is burned and the aircraft becomes lighter.

Lockheed made an attempt to optimize the existing L-1011 autothrottle system to the speed control tolerances sought for the flight management system, but the effort was abandoned. Lockheed engineers said autothrottle use under these conditions was characterized by excessive throttle activity that would have annoyed flight crews, disturbed passengers, increased maintenance costs and possibly compromised engine manufacturers' warranties.

Design objectives

For the TriStar flight management system the objectives were to control speed within ±0.005 Mach from the target Mach number in light turbulence (less in smooth air), hold maximum variations in engine pressure ratios to ±0.01 (±2% thrust) about the trim EPR, and avoid excessive throttle movements to preclude exceeding engine limits.

Concern over the autothrottle behavior eventually led to the adoption of a pitch control system for short-term speed adjustments, with power changes used for longer-term speed and altitude variations. If Mach begins bleeding off slowly, for example, a small nose-down command is applied by the autopilot to stop the deceleration. The autopilot altitude hold mode remains engaged.

If Mach begins increasing, on the other hand, a slight nose-up command is applied by the automatic pilot. Altitude excursions under these circumstances normally do not exceed ±50 ft., according to Lockheed engineers. The company believes this concept is unique to the TriStar flight management system and is the key to the tight tolerances the system is capable of holding.

On the *Aviation Week* evaluation flight, the system held the computed Mach numbers with negligible variation, and nose or throttle movements to stabilize speed were imperceptible during the flight. As the aircraft turned northeastward toward Los Angeles International Airport on the next to the last leg of the round-robin flight from Palmdale, the minimum-fuel cruise speed option was selected in place of the minimum-cost option. The flight management system slowly reduced the speed from Mach 0.834 to 0.804.

Like the minimum-cost mode, the minimum-fuel cruise mode uses aircraft gross weight, the selected cruise altitude and actual winds in conjunction with stored aircraft and engine performance data. From these, Lockheed technicians say, the flight management computer derives the optimum cruise Mach number for minimum fuel usage. Unlike the minimum-cost mode, however, fuel costs at the refueling site and hourly direct operating costs are not considered.

In addition to optimum speeds, the flight management system also computed optimum altitude continuously throughout the flight. The optimum altitude displayed on the flight management system's CRT changed as fuel was burned and the aircraft became lighter. The recommended altitude rose from 37,500 ft. at the beginning of the flight to 38,300 ft. about 1 hr. later.

In considering other options, such as step-climbing to a higher altitude, the system provided the wind limits at the new altitude which, if exceeded, would have made it economically unfeasible to climb to the higher altitude.

Options available

In addition to winds, the computation of optimum altitude took into account initial aircraft weight, destination airport elevation, temperature deviation from standard, selected climb ratings and speeds, predicted climb fuel usage and distances between flight plan waypoints. On shorter sectors, it can be more cost efficient to cruise at lower altitudes.

In addition to the minimum cost and minimum fuel cruise options available in the L-1011 flight management system, flight crews can select a maximum endurance airspeed computed by the system, or any particular Mach or indicated airspeed dictated by air traffic control or other circumstances. A speed of Mach 0.83 was selected on this flight while inbound to Los Angeles, and the speed built up smoothly over several minutes from our minimum fuel cruise speed of Mach 0.804.

A substantial amount of performance data is always on call from the flight management system. In response to a request for performance limits with one of the underwing engines inoperative, for example, the CRT responded with a net ceiling (the altitude at which the aircraft can still maintain a 1.4 deg. climb gradient) of 28,200 ft. under then-existing conditions.

Optimums indicated

It also was able to provide at that point the distance remaining on two engines before the aircraft began using its fuel reserves of 810 mi. and the estimated time en route to that point of 1 hr. 56 min. Additionally, it provided the optimum indicated airspeed and EPR for the remaining engines (221 kt. IAS and 0.669 EPR).

Though air traffic control requirements may make it difficult to take advantage of it, the system also provides an optimized descent profile and advises the pilot as he nears the begin-descent point. That point is determined by first feeding the computer the end-descent point, altitude and airspeed where the pilot would like to terminate the descent—at an initial approach fix, for example. The flight management system then considers aircraft performance parameters, cruise altitude and existing winds at cruise to compute the beginning of descent point. On this flight, the end-of-descent point was designated as a waypoint several miles south of Los Angeles International at an altitude of 5,000 ft. and an airspeed of 220 kts.

As in the climb and cruise phases, there were several options available for the descent. These included a minimum cost mode, minimum fuel mode, two selectable descent schedules of Mach 0.85 (350 kt.) IAS and Mach 0.82 (320 kt.) IAS, a rough air penetration schedule of Mach 0.82 (285 kt.) IAS, and a non-selectable schedule of Mach 0.80 (300 kt.) IAS to which the system automatically resorts if the aircraft is exceptionally light.

In initial tests of the TriStar flight management system, Lockheed found that the aircraft could not follow the theoretical optimum descent profiles comfortably, which sometimes were too steep. Eventually, the descent profile options were matched to the possibilities of the aircraft. Emphasis now is not on following a particular profile, but rather on maintaining a given airspeed and a given thrust.

Flight crew advised

Once the descent is begun, the flight management system tells the flight crew if the aircraft will arrive long or short of the end-of-descent point. Power can then be adjusted slightly to reach the point accurately.

On this flight, the flight management system CRT began flashing about 10 min. before arrival at our beginning-of-descent point, advising that the descent mode should be armed. This can be accomplished any time within 10 min. of the beginning of descent, and in this case was accomplished at 4.8 min. before reaching the beginning-of-descent point.

The flight management system automatically initiated the descent when the time arrived, pushing the nose down slowly and reducing power to establish our preselected descent schedule of Mach 0.82 (320 kt.) IAS. As the descent stabilized, the flight management system CRT indicated the aircraft would be 6-7 mi. short of the designated end-of-descent point.

EPR was increased slightly on the No. 2 engine, and the aircraft descended at 2,000 ft./min. Passing through 16,000 ft., the navigation system switched to a narrower search pattern and began looking for VOR stations within 80 naut. mi. Finding none, it switched back to Santa Barbara VOR to the north.

At 10,000 ft., the aircraft leveled off momentarily, slowed to 250 kt. to comply with FAA regulations, then resumed the descent. Over the designated end-of-descent point, the aircraft leveled off at 5,000 ft. (210 kt.) IAS.

New altitude of 15,500 ft. and 250 kt. was selected to overfly Los Angeles en route to Palmdale for a series of automatic approaches there. Like the rest of the flight, the approaches and landings were mostly automatic and provided a good demonstration of the system's capabilities under some difficult crosswind and wind shear conditions at Palmdale.

Landings were made on Runway 25, with a 30 kt. wind from 280 deg. This necessitated an approximate 6-deg. crab on approach, which was removed by the automatic landing system at 150 ft. altitude and replaced by a wing down/opposite rudder correction held by the system until touchdown. Despite the winds, there appeared to be no deviation from the runway centerline.

Approximately a quarter mile from the runway threshold there was a distinct wind shear of approximately 15-20 kt., which necessitated the addition of power at that point of each approach. On a manual approach flown by this editor using the aircraft's autothrottle system, the throttles went to nearly full power passing through the wind shear, then came back to a lower power setting as conditions stabilized. It was a clear demonstration of the system's ability to determine the degree of reaction needed in such a situation more quickly than the flight crew itself.

Prior to a final landing, the capabilities of the active aileron controls were also examined on the L-1011 prototype. At 17,500 ft. and 300 kt., a special computer in the cabin of the aircraft commanded small movements in the horizontal stabilizer that in turn induced flex loads and bending moments in the wings.

These movements were hardly perceptible within the aircraft, but a printout in the cabin of inputs from sensors in the wings showed significant excursions from a static wing profile. Turning on the active aileron control system damped out the excursions almost immediately, however, demonstrating the capabilities of the system in turbulent air.

Digital system no different

The system installed on the prototype was an analog system, but Lockheed engineers said there would be no significant difference in the behavior of the digital system that will be installed on operational L-1011-500s.

Testing is continuing, and the first operational active controls are expected to be installed in L-1011-500s that will be delivered in late 1980. Lockheed engineers estimate that the flight management system by itself will bring about a reduction of 2-5% in TriStar fuel consumption in commercial service, depending primarily on the length of flights.

The active aileron controls are expected to provide another 3% reduction in fuel used. Beyond this, Lockheed is still pursuing the possibility of applying active controls to the horizontal tail of the L-1011, permitting a reduction in the size and weight of the tail (AW&ST Sept. 19, 1977, p. 26). This, in turn, could reduce fuel consumption by another 3.5% on the TriStar.

757 systems key to route flexibility

Robert R. Ropelewski/Palmdale, Calif.
August 30, 1982

Light, responsive handling qualities of Boeing's new 757 commercial transport, combined with the aircraft's versatile flight management system and automatic flight controls, will offer airline crews increased flexibility and reduced workloads on the short- and medium-range routes the 757 is intended to serve.

This *Aviation Week & Space Technology* pilot flew the 757 on a 2-hr. evaluation flight within 24 hr. of flying the company's larger and recently certificated 767 (AW&ST Aug. 24, p. 40), and found the cockpit layouts and operating procedures of the two aircraft to be almost indistinguishable except for the lighter control forces and crisper control responses of the 757.

Boeing 757 cockpit, nearly identical in size and layout to that of Boeing's larger 767, relies extensively on digital flight management systems, cathode ray tube electronic displays, and microprocessor computer technology to limit the workload for two-crewmember operations.

The first 757 made its maiden flight on Feb. 19 (AW&ST Mar. 1, 1982 p. 34) from Renton, Wash., Municipal Airport, powered by two 37,400-lb.-thrust Rolls-Royce RB.211-535C engines. The aircraft is expected to be certificated in December for commercial airline service beginning early next year, and later will be available with either advanced versions of the RB. 211-535 or the 37,000-lb.-thrust Pratt & Whitney PW2037 engine.

The integrated flight management computer, autopilot/flight director, thrust management system and color cathode ray tube displays used in the 757 bring a new level of cockpit sophistication to the short- to medium-haul transport market, and should help ease crew workloads on multistop, short-haul route systems where workloads are especially high because of frequent takeoffs and landings. Boeing officials believe the commonality of the 757 cockpit with the 767 (AW&ST Aug. 9, 1982, p. 28) will enable flight crews to obtain a common type rating for both.

The 757 evaluation flight was flown from Palmdale Airport north of Los Angeles, where Boeing flight test crews were finishing a series of takeoff performance demonstrations necessary for Federal Aviation Administration certification of the aircraft. Boeing project test pilot John H. Armstrong occupied the right seat for the flight, which was conducted in the No. 2,757 flight test aircraft, the first to carry Eastern Airlines colors.

Although the 757 is a standard single aisle, narrow-body aircraft and the 767 is a twin-aisle, wide-body transport, Boeing adopted a blunter nose configuration for the 757 in 1979 (AW&ST Aug. 6, 1982, p. 22). This extended the fuselage constant cross section farther forward and thereby widened the cockpit by about 2 ft.

Second Boeing 757 flight test aircraft is backed into parking space at Lockheed California Co. flight test facility at Palmdale, Calif. This aircraft was flown on the Aviation Week & Space Technology *evaluation flight while at Palmdale for FAA certification demonstrations.*

As a result, the 757, which has the same fuselage cross section as the Boeing 727 medium-range trijet, carries a wide-body cockpit that is essentially the same size as that of the 767 from the pilot's seats forward. Forward windshields, instrument panels including electronic attitude, horizontal situation, and engine and system displays, center control pedestal and overhead panels are identical. Many of the avionics systems of the two aircraft types are identical and interchangeable as well, including air data computers, flight control computers, flight management system computers and control display units, communication and navigation radio control panels, inertial reference units and control panels, and thrust management computers and selector panels.

Over 80% of the line replaceable units in the flight management system are interchangeable between similar configurations of the 757 and 767, according to Boeing, and have identical part numbers. Components that have engine-related software, such as the thrust management and flight management computers, will be interchangeable between 757s and 767s powered by engines from the same manufacturer. In addition to simplifying crew training, this commonality is expected to reduce maintenance and support costs for airlines operating both aircraft. It also will increase opportunities for pooling aircraft and spares among airlines and reduce dispatch delays through the wider availability of spares.

Benefits of this cockpit commonality were evident on the flight in the 757, because the previous day's briefings in a cockpit procedures trainer at Boeing facilities in Seattle were equally applicable to both the 757 and 767. The transition from the 767 to the 757 was eased by the fact that the flight deck equipment and layout were identical, and most of the performance characteristics and reference speeds were very similar.

Operational changes

One of the few significant operational differences was aircraft weight. The 757 has a maximum takeoff weight of 220,000 lb., compared with a maximum takeoff weight of 300,000 lb. for initial 767s. The 757 I flew here at Palmdale weighed about 193,800 lb. at engine start.

As with the 767, the 757 cockpit is both comfortable and efficient. The forward instrument panel, with its extensive use of color electronic video displays for attitude, horizontal situation and engine and subsystems instrumentation, gives the impression of spaciousness on the flight deck. Because the aircraft was designed for two pilot operations, all essential switches and indicators are within sight and reach of either crewmember. The impression left after my flight in the 757 is that the aircraft can be operated without difficulty by a single crewmember, if necessary.

Simplified prestart checklist for the 757 follows a natural path from the overhead panel to the forward instrument panel and down the center control pedestal. The overhead panel is divided into five distinct vertical columns, facilitating an easy top-to-bottom scan of each column from left to right. Lighted pushbutton on/off switches are used extensively on the overhead panel, saving space and reducing the movement necessary compared with traditional toggle switches.

Other systems controlled from the overhead panel, such as demand hydraulic pumps, environmental controls and passenger seat belt/no smoking signs, have knobs with an "automatic" position and normally do not have to be touched during an entire crew cycle. Seat belt/no smoking signs, for example, are keyed in the automatic mode to landing gear retraction and extension (with appropriate delays for the takeoff phase). No crew action is necessary.

Many of the pressure, temperature and other indicators of system functions also have been eliminated from the overhead and front instrument panels and appear only on the engine indication and crew alerting system (EICAS) video displays on the center instrument panel when specifically summoned by the crew or when operational limits are being approached or exceeded. Throughout the startup of the 757 for my evaluation flight, we used only the engine pressure ratio (EPR) indicators for the two Rolls-Royce RB. 211-535 engines on the upper EICAS display and the engine high pressure compressor indicators on the lower EICAS display. Remaining checklist actions to be completed were indicated also on the EICAS displays until they were accomplished.

The flight management computer is an integral part of the total flight control system on the 757, and we loaded our flight plan—a round-robin route heading northward from Palmdale, then eastward toward Las Vegas, then southwesterly toward Palmdale—into the computer using the control display units on each side of the throttle quadrant on the center pedestal. Although the two computers can operate autonomously, the units communicate with one another, and all of the route information entered into one of the computers is transferred automatically to the other.

Geographic coordinates for all of the radio navigation fixes and airports along the proposed route were stored in the computers. It was necessary only to enter the letter designators for each fix to program my route in the flight management system.

Graphic layout

One of the helpful aspects of the 757's integrated systems and displays is that the electronic horizontal situation indicator (EHSI) can be used on the ground in the "plan" mode to lay out graphically and con-

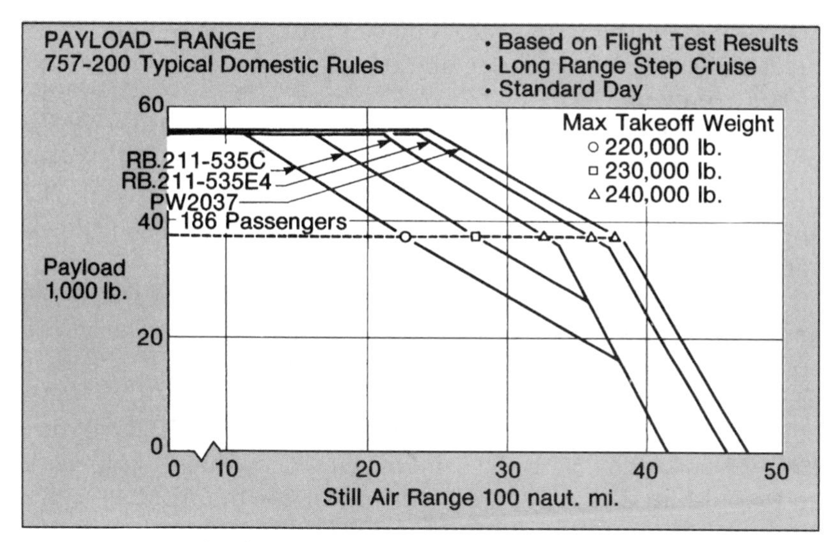

Boeing 757 payload-range chart illustrates the range capabilities the aircraft will have with the various engines that are planned for certification on the new commercial transport.

firm the flight plan route before entering it into the computer. Once the crew is satisfied with the proposed route and the "execute" button has been pushed on one of the flight management system control display units, the plan becomes the basis for the lateral navigation steering signals sent from the FMS to the autopilot/flight director.

Eventually, the flight management system will have a performance management and vertical navigation capability that will compute automatically the most economical climb, cruise and descent profiles, including fine-tuning chores such as wind/altitude tradeoffs and crew time/fuel burn economic assessments. Software for these functions is under development for the 757 and 767 flight management systems.

Without autopilot

Initial phase of my flight was flown without the autopilot. As we taxied onto the Palmdale runway, I advanced the throttles to the full forward position. The 757 uses electronic engine controls that inhibit the engines from exceeding their rpm or temperature limits, no matter where the throttles arc placed. There is a linear relationship between the throttle lever angle and engine thrust when the electronic engine controls are operating, and it is possible to display the commanded thrust levels on the EICAS screen immediately as the throttles are moved to any given position, well before the actual thrust stabilizes at those levels. The result is that engine thrust can be set quickly and precisely on takeoff or during any other phase of flight, freeing the crew's attention.

The weight of the 757 was approximately 193,000 lb. and our go/no-go decision speed and rotation speed were 129 kt. I rotated the nose at that speed and the aircraft lifted off after a takeoff roll of approximately 5,000 ft. Temperature was around 80°F. Palmdale elevation is about 2,500 ft.

I leveled off initially at 13,000 ft. to get the feel of the aircraft and examine its low speed handling characteristics. The aircraft could have climbed to a 40,000 ft. initial cruise altitude, according to Boeing's performance charts for the 757 with the RB. 211-535C engines. Initial cruise altitude would be about 1,000 ft. lower for a PW2037-powered 757 at the same weight, according to the charts, and would be a few hundred feet higher for the RB. 211-535E4-equipped 757.

Light control

At an altitude of 13,000 ft. and a speed of 300 kt., the 757 needed only light control forces to initiate relatively crisp roll rates left and right. Unlike many other transport aircraft on which outboard ailerons are desensitized or locked out as airspeed increases, the 757's ailerons re-

PERFORMANCE SUMMARY
757-200 Typical Domestic Rules

• 186 Passengers
• Based on Flight Test Results

Max Takeoff Weight lb.	220,000		
Engine Type	RB. 211-535C	RB. 211-535E4	PW2037
Design Weight lb.			
Max Landing Weight	198,000	198,000	198,000
Max Zero Fuel Weight	184,000	184,000	184,000
Operating Empty Weight	129,100	129,350	128,600
Design Range naut. mi.	2,260	2,510	2,650
From Denver (84° F) naut. mi.	2,260	2,510	2,650
From Denver (92° F) naut. mi.	1,940	2,430	2,470
Altitude Capability ft.			
Cruise Ceiling (MC$_R$T, max wt.)	38,400	38,700	37,700
Engine Out (1,000 naut. mi. mission)	18,100	19,500	20,200
Takeoff and Landing			
TOFL (SL, 84° F at max wt.) ft.	6,590	6,530	6,490
Approach Speed, KEAS			
Mission Landing Weight	127	126	126
Maximum Landing Weight	134	134	134
Block Fuel/Seat, lb./Seat			
500 naut. mi.	59.1	53.2	52.0
1,000 naut. mi.	103.8	94.1	91.6

Boeing 757 performance summary, based on flight test results to date, show that subsequent versions of the Rolls-Royce RB.211-535 engine and the new Pratt & Whitney PW2037 powerplant planned for use on the 757 will give the aircraft better fuel efficiency and range.

main effective at all speeds. The aircraft has a demonstrated roll rate of 90 deg. per sec. with full lateral deflection of the control wheel. A given roll rate can be achieved with less wheel deflection than would have been the case on earlier Boeing aircraft.

Roll rates equivalent to the 727 trijet, for example, are achieved in the 757 with less wheel deflection. Pitch forces also gave the impression of being lighter than other transports in the Boeing family. The 757's dual channel pitch augmentation control system provides a positive elevator control force gradient that increases wheel forces according to the rate and force of pilot inputs. The easier I pulled, the lighter the forces; the harder and faster I pulled, the heavier the pitch forces became.

Still at 13,000 ft., I pulled the throttles back and activated the spoilers to slow the aircraft for a series of stalls. Spoiler extension at high speeds causes a barely noticeable pitch up that was countered easily with nose-down trim using the trim switches on the control wheel. Another deviation from previous Boeing cockpits is the absence of the large manual pitch trim wheels located on the sides of the center pedestal in the 707, 727, 737 and 747.

With the aircraft in the clean configuration, I tried to keep our deceleration rate as close as possible to 1 kt. per sec. until the stall. Stall speed was predictably high for the clean aircraft. The wheel shaker warning system began vibrating the control wheel as the airspeed dropped below 160 kt.

The stall itself, accompanied by a sharp increase in our rate of descent, occurred around 150 kt. The aircraft showed no tendency to roll off on either wing and ailerons and rudder remained effective. The nose pitched down slightly, even with the wheel held at its aft limit, and the aircraft stabilized in that attitude with a noticeably strong aerodynamic buffet shaking the airframe until I pushed the nose down a few more degrees and increased power to accelerate out of the stall.

Similar characteristics were exhibited by the 757 in all flap configurations. With full landing flaps and landing gear extended, the wheel shaker activated at around 110 kt. and the full stall occurred at about 100 kt. We continued our climb to observe the aircraft at some typical cruise altitudes.

Fuel burn

Passing through 25,000 ft. at 260 kt. with a climb power setting of 1.61 EPR on the Rolls-Royce engines, the aircraft was burning fuel at a rate of 6,700 lb./hr./engine, according to the fuel flow indicators we selected for display on the lower EICAS display.

A left rear entry door caution light appeared on the left side of the upper EICAS display about this time, but a check of the door showed it was firmly closed and locked. The problem was later found to be a faulty microswitch in the door.

Another caution indication, this one for the left engine high-pressure compressor, appeared as we established the aircraft in cruise at 29,000 ft. This turned out to be an engine bleed valve malfunction that occurred only in a very narrow range of the engine's operation, and it did not affect our flight. Both of the caution messages were removed from the EICAS screen by pushing a "cancel" button to the left of the upper EICAS display. Armstrong said there have been few false caution or warning messages associated with primary systems since 757 flight testing began, a fact that has surprised many associated with the program.

Although the 757 is barely halfway through its planned schedule, the electronic displays appeared to be working flawlessly. There was no flickering or blanking out of any of the flight or systems displays during the flight, and bright sunlight shining into the cockpit did not seem to diminish the distinctiveness or readability of the displays.

Course tracking

While we were conducting our stalls, then climbing and leveling off at various altitudes, the flight management system continued to track our course using the aircraft's three Honeywell inertial reference units and the automatic very high frequency omnirange (VOR) and distance measuring equipment (DME) tuning capability of the flight management computer.

Our position with reference to our flight plan route was displayed continuously on the electronic horizontal situation indicators (EHSIs). This diminished the need to take a break from our other activities for a periodic position fix.

As we approached en route way points where turns were required to a new outbound beading, a flight path trend vector—a dashed line ahead of the aircraft symbol on the EHSI—showed the aircraft's predicted flight path for any given angle of bank and turn rate. By starting our turn two or three miles before the way point and rolling the aircraft until the outbound course line on the EHSI was tangential to the curved trend vector, it was possible to effect a smooth and precise intercept of the outbound course without a considerable amount of guesswork or hunting to bracket the new course. It would have been even easier, of course, with the autopilot engaged, in which case the flight management computer would have handled the task automatically.

The relatively low fuel consumption of the Rolls-Royce engines was noteworthy at all altitudes during the cruise portion of our flight. In level flight at 33,000 ft., with a speed of 0.8 Mach/285 kt. and a thrust setting of 1.42 EPR on both engines, our fuel flow indicated 3,900 lb./hr./engine on the EICAS.

Armstrong said that 0.8 to 0.81 Mach is the most economical speed range for the 757 wing. Trying to maintain a slightly higher speed increased fuel flow significantly. In level cruise at 35,000 ft., 0.85 Mach/292 kt., with engine pressure ratios at 1.62, our fuel flow was 5,300 lb./hr./engine.

I engaged one of the three Sperry autopilots briefly at that point and it held our heading, airspeed and altitude precisely until I disengaged it again several minutes later.

Rather than continue our flight as far east as Las Vegas, we amended the flight plan shortly to return to Palmdale and conduct automatic and manual approaches and landings in the 757. Capabilities of the flight management system and EHSI facilitated the change in the flight plan route. While I continued to use my EHSI to fly the clearance flight plan, Armstrong selected the "plan" mode on his EHSI to display a new routing for our return to Palmdale. The new routing was established using the flight management computer on his side of the center control pedestal.

Waypoint display

Once this was finished, we contacted Los Angeles center to file our amended flight plan. When it was approved, Armstrong pushed the "execute" button on his flight management control display unit to enter the new flight plan route into the flight management system. As he did this, the old route disappeared from my EHSI and the new route to the next waypoint appeared. Distance and time readouts at the upper left and upper right corners of the EHSI indicated how far away the waypoint was and how long it would take to get there at our present ground speed. An available option displays estimated time of arrival expressed in Greenwich Mean Time in place of time remaining to the next waypoint.

Armstrong had demonstrated single engine flight characteristics of the 757 successfully on the first flight of the 757 on Feb. 19, when the No. 2 engine was shut down and restarted after a compressor stall occurred (AW&ST, Mar. 1, 1982, p. 34).

EICAS capability to reduce crew workload was demonstrated then as well. An in-flight engine restart is a source of significant workload on a twin jet transport with high-bypass engines because these engines have a more restrictive relight envelope than the low-bypass engines on previous twin jet aircraft.

When an engine on the 757 must be restarted in flight, the three closest flight levels for a successful relight at or below the present altitude of the aircraft are displayed in 2,000 ft. intervals on the EICAS, along with the required airspeed envelope. The proper wind milling speed for each rotor is indicated as a cursor on the rotor speed displays. The instruction to use crossbleed is also specified, if required. On previous aircraft, this data had to be determined by the crew from charts in the operations manual.

Cruise altitude

As we neared Palmdale, approach controllers were unable to clear us for a descent from our cruise altitude until it was much too late for a normal letdown. I reduced the throttles to idle, extended the speed brakes, and Armstrong extended the landing gear in order to increase drag and expedite our descent. Even under these circumstances, cockpit noise levels remained low enough to talk in a normal tone of voice between the two seats. This had also been the case in the cruise portion of our flight, and there was little noticeable difference in noise levels from low speed up to the 757's maximum operating speed of 0.86 Mach.

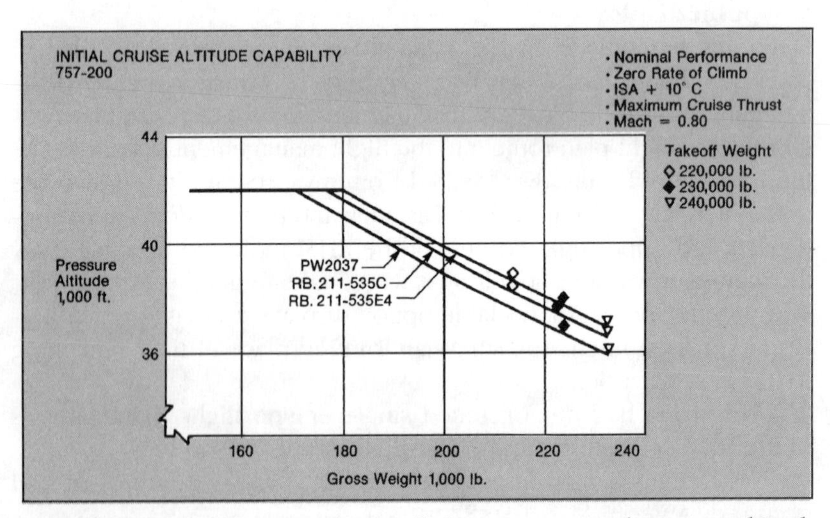

Boeing 757 twinjet transport will have an initial cruise altitude capability in the 36,000-39,000-ft. range, depending on the engines used. The 757 will be certificated initially at 220,000 lb. takeoff weight, but weight increases to 230,000 lb. and 240,000 lb. are envisioned.

Our excessive altitude, some light aircraft traffic, occasionally heavy thermal turbulence, and the requirement to avoid an unseen Lockheed U-2 that was also in the Palmdale landing pattern added some complexity to our initial approach to Palmdale, but underscored the easy handling characteristics of the 757 in both the manual and automatic modes. I engaged the autopilot briefly to intercept the instrument landing system localizer and glideslope for Runway 25 at Palmdale, then disengaged the autopilot to fly the first approach manually. Outside air temperature was approximately 90F at that point, and a 7-10 kt. right crosswind was blowing across the 2,500 ft. elevation field from the north.

Under these circumstances, I found the 757 to be one of the few large transport aircraft that can be handled comfortably with a single hand on the control wheel and the other on the throttles, even with the thermal activity and crosswinds we were experiencing at the time. Whether under manual or automatic control, the aircraft responded quickly and precisely to control inputs, resulting in smooth, centerline landings on each of the several approaches we made. An approach reference speed of 126 kt. indicated was used for our initial landings.

Like the 767, the large windshield and windows and relatively low nose attitude of the 757 create the impression of being in an almost flat attitude during the approach. Visibility was excellent because of this, even during the flare. Raising the nose an additional 2-3 deg. seemed suitable for the flare, and resulted in a relatively soft touchdown on each landing.

On the next to last landing performed on this flight, Armstrong pulled the right throttle to idle just as the aircraft lifted off from a touch-and-go landing. A symmetric thrust was countered easily before the nose could yaw and the right wing could dip more than a few degrees. The aircraft continued to climb about 700 ft. per min. at a speed of 130 kt., with 20 deg. flaps and a 1.54 EPR setting on the left engine. The right engine was left at idle for our final approach and landing, using a reference speed of 147 kt. with flaps still at 20 deg. Aircraft weight was approximately 172,500 lb. Again, the aircraft was flown comfortably under those circumstances, even when a sudden loss of airspeed on short final necessitated a substantial increase in power on the left engine.

Reverse thrust was applied only on the left engine after touchdown. The aircraft was kept on the centerline using left rudder and brake and nose wheel steering until we were slow enough to taxi off the runway and return to the flight line.

January delivery

Initial deliveries of the 757 are scheduled to begin in January, 1983, to Eastern Airlines. Eastern has ordered 27 of the aircraft and holds options on another 24. The early Eastern aircraft will be in essentially the same configuration seen on my evaluation flight—powered by 37,400-lb.-thrust RB. 211-535C engines and with maximum takeoff weight of 220,000 lb.

Deliveries of a 230,000-lb. takeoff weight 757 powered by 40,100-lb.-thrust RB. 211-535E4 engines are scheduled to begin in January, 1984. This will involve no increase in the empty weight of the 757. An even heavier 240,000-lb. takeoff weight version of the 757 powered by 38,200-lb.-thrust Pratt & Whitney PW2037 engines is scheduled to be available for delivery in February, 1984. Empty weight of this version will be about 100 lb. heavier than the basic aircraft.

Numerous growth provisions are included in the basic aircraft, particularly in the flight management system. Software for performance management and vertical navigation features are in the validation phase, and will allow 757 crews new capabilities to determine climb and descent speeds and power settings, top of descent points, and other economy-related data.

Four-dimensional navigation

Other growth functions envisioned for the flight management system include four-dimensional navigation allowing the aircraft to arrive at an initial approach point within seconds of a prefiled time. Omega low-frequency radio navigation, microwave landing system, Navstar Global Positioning System navigation inputs, collision avoidance system, wind shear detection, actual versus data base aircraft performance monitoring, noise abatement flight procedures and remote automatic flight plan loading in the flight management computer. Much of this will involve only software changes in the present flight management system, and will not entail a substantial amount of additional hardware.

Carbon brakes that are expected to be available for delivery or to be retrofitted on the 757 in the fourth quarter of 1984 will allow an estimated weight reduction of 650 lb. while lasting up to five times longer than conventional steel brakes.

Boeing's new 767 eases crew workload

Robert R. Ropelewski/Seattle
August 23, 1982

A significant increase in cockpit automation and a corresponding decrease in traditional cockpit workloads, rather than any dramatic improvements in performance and handling qualities, will be evident to airline crews operating Boeing's new 767 transport scheduled to enter commercial service beginning next month.

The emphasis Boeing placed on procedural simplification and operational ease was observed first-hand by this *Aviation Week & Space Technology* pilot during a 1-hr. 45-min. demonstration flight in the 767 shortly after it was granted FAA type certification.

The 211-255 passenger, twin-aisle, wide-body 767, which was certificated July 30 by the Federal Aviation Administration (AW&ST Aug. 9, 1982, p. 28), is being delivered to United Airlines for inauguration into scheduled service Sept. 8. Extensive use of versatile and reliable digital avionics throughout the 767 cockpit is expected to ease the workload of the two-man crews who will be flying the aircraft. Emphasis clearly has been put on a systems management rather than a hands-on approach to the operation of the 767.

The aircraft flown was the seventh 767 to come off the production line. It is the first to have the two-man cockpit and will be the fifth 767 to be delivered to United. This aircraft was used to conduct initial demonstration flights by Boeing and United Airlines pilots between Seattle and Chicago in June (AW&ST June 21, 1982, p. 28) and subsequently was used on a European demonstration tour.

Apart from its high level of automation, the 767 has retained the stability and handling qualities of earlier Boeing transports. Control forces are slightly on the heavy side, though responsiveness has been improved from previous aircraft such as the 707, especially at low speeds. The new aircraft also has very straightforward stall characteristics, with natural and artificial stall warning margins adequate enough to prevent inadvertent stalls under most circumstances, These features were all examined during the *Aviation Week & Space Technology* evaluation flight.

Spaciousness of the 767's cockpit is similar to that of other wide-body transports currently in service, but the absence of a flight engineer's panel at the rear of the flight dock makes the cockpit feel even roomier. Late adoption of the two-crew configuration in the evolution

Engine indication and crew alerting system (EICAS) cathode ray tube displays on center instrument panel of the Boeing 767 can be used to show engine and subsystem operating parameters, and also provide the crew with warning and advisory information on aircraft subsystem malfunctions. Normally, the bottom screen remains blank until the crew specifically requests an engine or systems status check. Engine pressure ratios (EPR), low-pressure compressor speeds (N_1) and exhaust gas temperature normally are displayed continuously on the top screen, but all information can be projected on either screen. The small panel at upper right of the top EICAS display is the thrust computer panel, which enables crewmembers to command the autothrottle to compute and set takeoff/go-around power, full or derated climb power, maximum continuous thrust or cruise thrust. Flap position indicator and alternate flap selector knob are at center right. Automatic brake selector knob is at lower left, with standby digital engine instruments panel.

of the 767 design has resulted in a greater amount of space at the rear of the flight deck than would otherwise have been the case.

To accommodate all necessary switches and gauges within the reach of the captain's and first officer's seats, Boeing has automated the operation and monitoring of many of the aircraft's systems. Typical of these is the auxiliary power unit, which consists of a single three-position knob on the overhead panel above the captain. Turning the knob from "off" to "start" initiates the start process, and the rest is automatic. There are no gauges to monitor because the system monitors itself and automatically shuts down if any operating parameters are exceeded. Otherwise, an "on" light illuminates on the overhead APU panel when the system is ready. If desired, APU operating parameters can be displayed on one of the cathode ray tube displays on the center instrument panel.

Dominant characteristic of the 767 cockpit is the extensive use of cathode ray tube displays in conjunction with the Rockwell-Collins engine indication and crew alerting system (EICAS) and electronic flight instrument system (EFIS), and the Sperry flight management system (FMS). Each pilot has a multicolor electronic attitude director indicator and electronic horizontal situation indicator on his instrument panel as well as his own flight management system control display unit at the forward end of the center control pedestal.

Two additional 6-×-7-in. multicolor electronic displays associated with the engine indication and crew alerting system are mounted in the center instrument panel, and an automatic flight control system mode control panel is located in the instrument panel glare shield, in an arrangement similar to that of other recent-generation commercial transports.

Although most of the flight information needed by the crew is presented on the electronic displays, backup electromechanical instruments still fill most of the rest of the space on the instrument panel. As a result, the panel does not give the impression of being less cluttered than that of any other aircraft.

One of the methods used by Boeing engineers to reduce the cockpit workload in the 767 was to leave as many systems as possible in the permanent "on" condition so that control panels do not have to be scanned repeatedly before and after APU and engine start. As a result, the prestart checklist is very short. This is somewhat deceptive, however, because the entire overhead panel is a single item on the checklist, even though the crew must scan individual switch positions in a top-to-bottom, left-to-right sweep of the panel during prestart and post-start checks. Most of the switches are the push-button type that give a lighted "on" or "off" indication when electricity is applied.

Displays appear

Once the APU is started, control knobs for the 767's three inertial reference systems were turned to the "align" position. We waited several minutes before the systems were fully operational. When the inertial systems were ready, the electronic attitude and horizontal situation displays appeared on the two cathode ray tubes on each side of the instrument panel. Once the inertial systems were ready, they were put into the "navigation" mode, which provides an inertial navigation

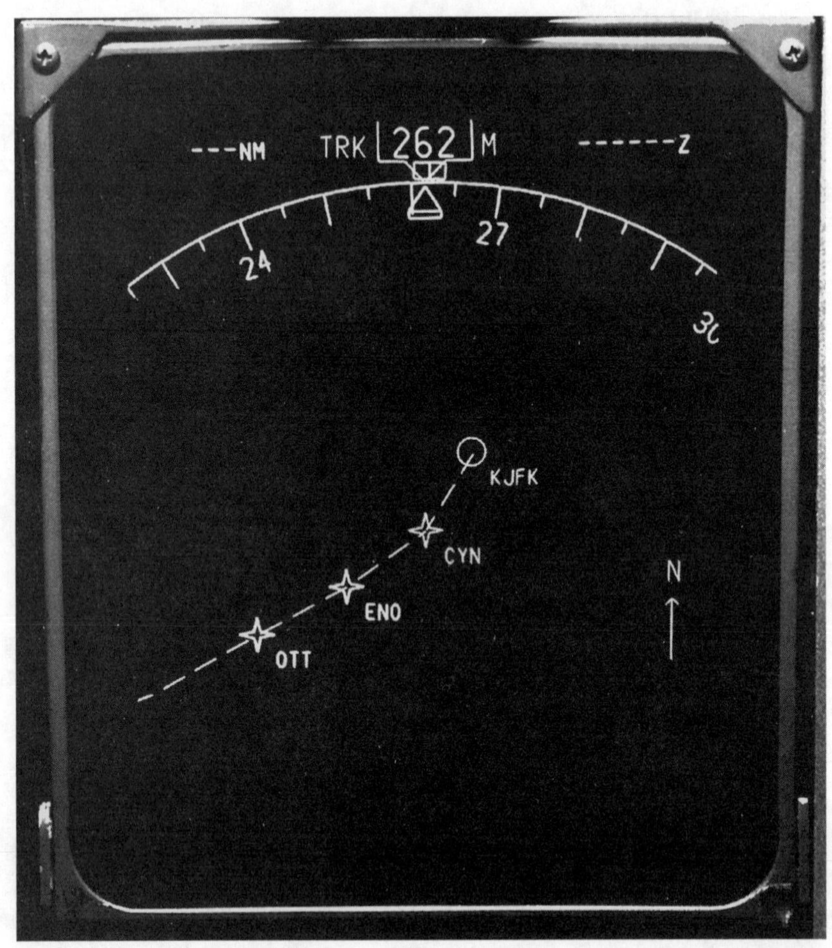

Electronic horizontal situation displays (EHSI) on the 767 instrument panel can be used in conjuction with the aircraft's flight management system to plan routes prior to takeoff. Small blue circle represents the departure airport, while the blue dashed line represents the aircraft's proposed flight path. Small crosses indicate position of departure route waypoints. When the route is confirmed and entered into the flight management system computer, the dashed line becomes solid.

capability as well as flight deck displays of magnetic and true heading and track, aircraft attitude, ground and wind speed and direction.

Operation of the flight management system computer was relatively easy and straightforward. It took only a few minutes to load into the computer our flight plan route from Paine Field in Everett, Wash., where

Boeing's 767 production facility is located, to Grant County Airport in Moses Lake, Wash. Software is still being developed for the vertical navigation mode of the flight management system, and only the standard lateral navigation capability was available on this particular flight.

Navigational fixes for instrument departures from the Seattle area were already stored in the flight management system computer and could be readily displayed on the electronic horizontal situation indicator as we received our clearance. In airline operations, it will be possible to store a large number of "canned" departure, en route and arrival plans in the computer so the crew will not have to take the time to program manually each phase of the flight.

Starting the two 48,000-lb.-thrust Pratt & Whitney JT9D-R4D engines on the 767 was as easy as starting the APU. With engine operating parameters displayed on the two EICAS video screens on the center instrument panel, engine start knobs on the overhead panel were turned one at a time from the off to the "ground" position. The start process was automatic from that point, and all that was necessary was to open the fuel valve for each engine when engine high-pressure compressor speed (N_1) reached 20%. Start procedures will be about the same for the 48,000-lb. thrust General Electric CF6-80A engines that will also be available on the 767.

One of the items not on the checklist was APU shutoff after engine start, and as a result, the APU was inadvertently left on for the entire flight. This posed no problem because the unit is cleared for operation throughout the 767 flight envelope.

One noteworthy aspect of the prestart and pretaxi procedures was the absence of the large number of aural warning tones often found in many recent transport aircraft. Boeing calls its 767 and 757 flight deck layout a "quiet, dark cockpit" in which indications of systems operations are reserved for conditions that require action by the flight crew. Aural warnings have been deemphasized. A reliance has been placed instead on the versatility of the EICAS displays to keep the crew advised on engine and systems status.

Many of the temperature and pressure indications that have long been standard in aircraft cockpits have been relegated to a standby mode in the EICAS, and they appear on the electronic displays only on request or when an out-of-tolerance condition exists.

Our aircraft weight was approximately 244,000 lb. as we took the runway at Paine Field for our initial takeoff. Boeing engineering test pilot James C. Loesch occupied the right seat. Maximum takeoff weight of the 767 is about 300,000 lb.

Throttles advanced

The throttles were advanced manually to establish an engine pressure ratio (EPR) of approximately 1.2 before engaging the automatic throttle mode by pushing the EPR button on the thrust management panel. The throttles then continued moving forward on their own to the EPR takeoff setting calculated by the autothrottle computer. One of the helpful aspects of the autothrottle system, in terms of workload and engine wear, is that the electronic engine controls will not permit engine parameters to be exceeded no matter how far forward the pilot pushes the throttles.

Rotation speed was 119 kt. with flaps in the 20-deg. takeoff setting, and as we reached that point, the nose was raised about 18 deg. above the horizon—a fairly typical attitude for most modern jet transports. Gear and flaps were raised as we climbed out, and at an altitude of 1,000 ft. the climb power setting was selected by pushing a button on the thrust management computer near the top of the center instrument panel. The computer calculated the optimum climb thrust for the aircraft's weight and ambient conditions, and automatically set the throttles to that thrust level. At the same time, alphanumeric readouts at the top of the upper EICAS video display showed that climb power had been selected and also indicated the EPR calculated by the thrust computer for the climbout. Small green carets on the electronic EPR gauges also moved to the proper value at the outside of the gauges to show where the EPR needles should be after the throttles were adjusted.

In both the autothrottle and manual throttle modes, small white bands appeared on the outside of the video EPR gauges each time the throttles were moved to show the momentary difference between the actual thrust level and the commanded thrust level, calculated by the electronic engine controls from the position of the throttles.

When the pilot operates the throttles manually, this system helps him set the throttles precisely and avoid exceeding the target thrust. Digital readouts of the actual thrust and target thrust were displayed above the video gauges on the EICAS.

Passing through 10,000 ft. on our departure from the Seattle area, the aircraft was climbing at a rate of approximately 3,200 fpm in climb power. Approaching 12,000 ft., Loesch pulled the right throttle to idle, simulating a failure of the right engine. The right wing dropped slightly until corrective left rudder and aileron were applied, and the aircraft continued to climb at a reduced rate of about 1,500 fpm with only the left engine operating at full power. Performance charts for the 767 indicate a single-engine altitude capability of about 25,000 ft. for an aircraft weighing 240,000 lb., the approximate weight of our aircraft at that point.

Boeing 767 in United Airlines color is towed toward a parking space on the 767 flight line at Boeing Field in Seattle, where it was returned following the Aviation Week & Space Technology *evaluation flight in the aircraft. Another 767 flight test aircraft, as well as additional 767s for United and a Delta Air Lines 767—the first to be equipped with 50,000-lb. General Electric CF6-80A engines—are in the background.*

Using the aircraft's distance measuring equipment, VHF omnirange (VOR) and inertial reference systems, the flight management system tracked our position as we were vectored out of the Seattle area to our flight plan route. The position, as well as heading, track, flight plan route, location of navigational aids, and distance and time to succeeding waypoints, was continuously displayed on the electronic horizontal situation indicators.

The flight management computer has a navigation data base capable of storing an airline's entire route structure, including both current data and the next revision, along with published terminal area procedures. This eliminated the need to keep a lapful of en route charts, terminal area charts and approach charts handy, because all of these were easily summoned on the EHSI. The flight management computer also has an automatic VOR/DME feature that continuously tunes the aircraft's navigational receivers to the appropriate navaids as the flight progresses.

Another time-saving and work-reducing feature of the flight management system is its ability to compute top-of-climb or bottom-of-descent points. Air traffic control considerations often dictate that an aircraft reach a newly assigned altitude by a specific geographic point, and crews generally have to compute manually an appropriate climb/descent rate to meet that requirement. The 767's flight management computer does this automatically during altitude changes, and displays a green arc on the EHSI to indicate where the aircraft will reach its new altitude at its current ground speed and rate of climb or descent. If the arc falls beyond the navigational fix at which the new altitude must be reached, for example, a thrust and/or pitch adjustment can be made to increase the climb or steepen the descent. Such adjustments are reflected immediately on the HSI with the movement of the green arc representing the target altitude.

Stability and handling characteristics of the 767 in the manual mode of operation made the aircraft easy and comfortable to fly at all speeds and altitudes without the automatic systems engaged. Crews transitioning into the 767 from other Boeing equipment should find the control forces and low-speed stability similar to those of the aircraft they are currently flying.

The aircraft also stalls in a straightforward manner. We conducted several approaches to stalls from 13,000-15,000 ft. on the outbound leg of our flight, and some actual stalls around the same altitudes on our return flight. Outbound, the stalling speed of the 767 was approximately 102 kt. at its roughly 240,000-lb. weight at that point, with a 20-deg. flap setting and landing gear up. Boeing has incorporated a stick shaker in the 767, and this began vibrating the control wheel at just above 110 kt. as the aircraft slowed. Natural buffet could be felt by about 105-106 kt., and power was applied at that point to stop the deceleration and recover.

On our return flight, the aircraft was allowed to continue decelerating until it stalled. This was done with landing flaps and landing gear extended and the aircraft at a much lower weight. In that case, I had to hold the control wheel at its full aft limit to reach the stall, at which point the nose dropped slightly and our altimeter began to unwind. Ailerons and rudder remained effective, and there was no tendency to roll in either direction. Re-applying power and lowering the nose just a few degrees more brought the aircraft out of the stall in a few hundred feet.

We eventually climbed to 37,000 ft. en route to Moses Lake. The aircraft was smooth and quiet at that altitude, even at its maximum operating speed (M_{mo}) of 0.86 Mach. Pitch stability of the 767 made a Mach trim system unnecessary, and none is installed in the aircraft. Likewise, there appeared to be little difference in the aircraft's behavior with or without yaw dampers. Several times during the flight it was necessary to add some right rudder trim, but Loesch said this was a peculiarity of the particular aircraft, which he said was poorly trimmed.

Low noise

Noise levels were noticeably low in the cockpit and the passenger cabin during the high-speed cruise portion of our flight, and it was

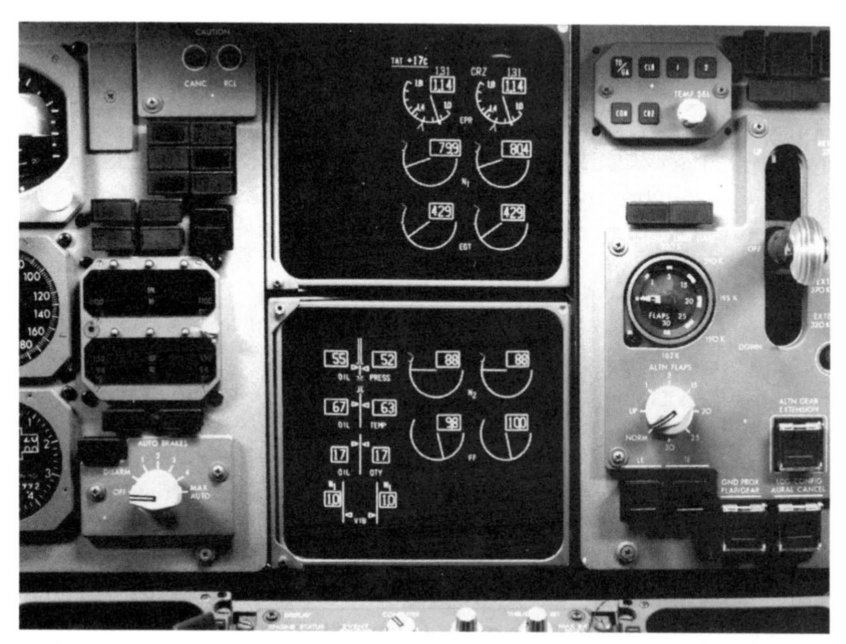

Electronic attitude director indicator (top) and electronic horizontal situation indicator (bottom) used in Boeing 767 and 757 transports provide crewmembers with a versatile and thorough display of attitude and navigation information without excessive clutter. Ground speed is displayed at upper left of EADI, while selected autopilot/flight director modes are indicated in green at bottom of EADI. Information at top of EHSI includes, from left, distance to next waypoint, magnetic track and estimated time of arrival at next waypoint. Triangular symbol near bottom of the EHSI marks the aircraft's present position, while a magenta-colored cross identifies the next waypoint and white crosses represent subsequent waypoints. Weather radar information is superimposed on the flight path display.

possible to converse normally in all sections of the aircraft throughout the flight. Boeing's own measurements of 767 flight deck sound levels, taken to insure that customer guarantees have been met, indicate that air-conditioning and aerodynamic noise levels are significantly below those of conventional standard-body transports and several decibels below other wide-body aircraft.

The aircraft also was relatively vibration free when speed brakes were extended to slow the aircraft and begin our descent toward Grant County Airport at Moses Lake. Buffet from the speed brakes was almost indistinguishable in the cockpit and forward and center sections of the cabin, and was only mildly discernible in the rear of the cabin where a large galley was located.

A feature not installed on the demonstration 767 was the vertical navigation mode in the aircraft's flight management computer. Software for this was still being validated but was expected to be incorporated in United's 767 by the time those aircraft go into service. With the vertical navigational mode operational, the flight management computer will be able to determine the optimum start-of-descent point and descent rate from cruise altitude to initial approach altitude or any other altitude selected in advance by the pilot. The system also would initiate the descent automatically.

Clear weather prevailed during our flight, and there was no need to use the aircraft's weather radar. In adverse weather the color weather radar can be displayed, properly oriented and scaled, on the electronic horizontal situation display. Route, navaid, airport and all other information normally available on the EHSI can then be superimposed over the weather imagery to allow crews to avoid severe weather areas and still remain oriented with respect to their flight plan route.

Had it been necessary to hold before beginning our approach to Moses Lake, we could have entered the pattern automatically after filing the holding fix position in the flight management computer. The flight management system has the capability of determining the proper procedure for entering the holding pattern, as well as for providing the necessary guidance to the autopilot to enter and maintain the pattern.

As we neared Moses Lake and were advised of the landing runway at Grant County Airport, it was also possible to enter this information into the flight management system and select an instrument landing system approach to Runway 32 at the airport, thus completing the flight plan in the flight management computer. Once this was done, one of the 767's three Collins digital autopilots was engaged on the flight control system mode control panel in the glare shield. This allowed the flight management computer to interact with the autopilot and steer the aircraft toward an intercept with the ILS localizer and glideslope.

A decision height of 200 ft. was selected and the desired altitude was displayed in the upper right-hand corner of the electronic attitude director indicator (EADI), just above a digital readout of our radio altitude. Ground speed is displayed continuously at the upper left corner of the EADI, while the lower corners of the display indicated in abbreviated form those modes of the automatic flight controls that were active, such as autothrottles, airspeed and altitude hold, localizer and glideslope capture, and flare and rollout. Localizer and glideslope deviation indicators and a fast/slow scale also are active on the EADI during ILS approaches.

The flight management system can acquire and fly the front or back ILS course to a runway. The three redundant autopilot channels provide a fail operational system that has been certificated for automatic flare, landing and rollout in Category 3B weather (zero decision height and 150 ft. runway visibility). The system flew a smooth intercept of the localizer and glideslope, and gear and flaps were extended for landing. A reference speed of 128 kt. was selected for the approach.

Beam distorted

The ILS at Grant County is not a Category 3 system, and this was evident when a Japan Air Lines Boeing 747 taxied onto the runway at the start of a training flight, distorting the ILS glideslope beam in the process. The nose of the 767 pitched up several degrees, then pitched down several degrees as the throttles advanced and then receded before finally stabilizing. These excursions were within the limits of the systems, however, and the aircraft continued its approach to a smooth touchdown.

Several more landings were conducted, including manual approaches and manual approaches with autothrottles engaged. The 767 has a relatively low approach speed about 126 kt. at typical landing weights and 136 kt. at maximum landing flown at speeds of 126-128 kt.

One favorable aspect of the aircraft in the landing pattern is the excellent visibility it provides of the runway environment. Front windshields are considerably larger than previous Boeing transports, creating the impression in the cockpit that the aircraft has an almost flat attitude on the approach that allows good downward visibility just in front of the aircraft. Loesch estimated that the average attitude is probably about 2.5-3-deg. nose-up for most 767 approaches. This will allow crews to see more runway and centerline lights for reference during landings in Category 3 weather conditions.

As the aircraft lifted off the runway following our final touch-and-go landing at Moses Lake, Loesch pulled the right throttle to idle to simulate another engine failure. I was still flying manually at that point but could easily counter the resulting asymmetric yaw with left aileron and rudder. With the landing gear and flaps raised, the aircraft climbed initially at about 1,700 fpm with a 10-deg. nose-up attitude. Trimmed, the aircraft held that attitude and continued the climbout.

The flight management system worked equally as well on the return to Everett, displaying aircraft position and flight path information, providing alternative routes to Everett, and generally keeping us

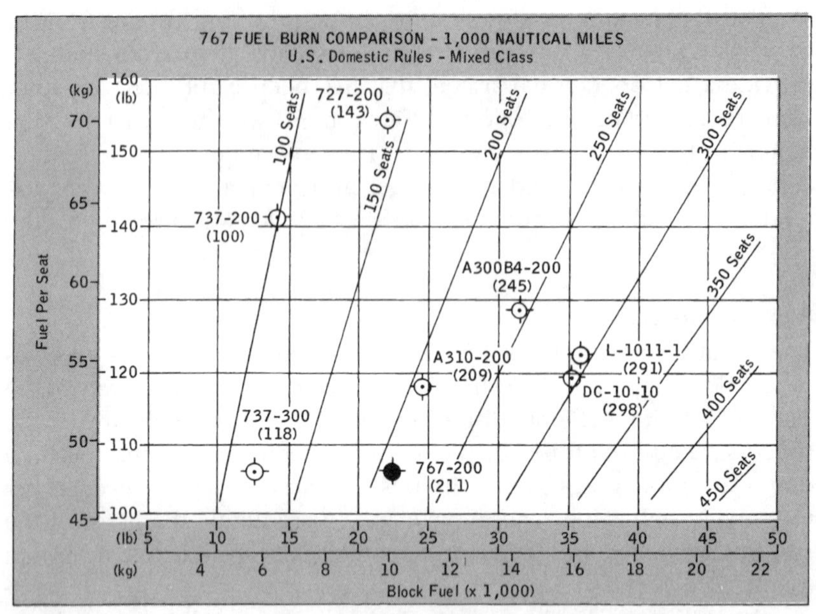

Fuel burn comparison pits the 767-200 against other Boeing transports and Airbus Industrie, Lockheed and McDonnell Douglas competitors. The comparison is made under U.S. domestic rules and for mixed-class passenger configurations.

more abreast of our progress than would have been possible without the system for those of us unfamiliar with the area.

With both engines operating normally again, another fully automatic approach and landing was executed to Paine Field. Because we were approaching the ILS localizer at Everett from more than a 90-deg. angle, we used the lateral navigation mode on the automatic flight control system to steer the aircraft to a point a few miles beyond the ILS outer marker, then turned the aircraft until it began to receive, and finally captured the localizer beam inbound. Once on the glideslope with the landing gear and flaps extended, we armed the brake system and spoilers for automatic activation on touchdown.

The automatic brake system selector knob on the lower section of the center instrument panel has several optional positions from 1 to 4, indicating increasing levels of braking intensity, and a final position labeled "max. auto." I selected the No. 2 position for our landing.

The smooth automatic approach and flare were terminated with a relatively hard touchdown at Everett, and the nose gear also came down with a surprising firmness as the wing spoilers deployed and automatic wheel braking began. Loesch attributed the hard landing to a slight left drift that developed as a crosswind correction was taken out by the autopilots just before touchdown.

Combination of spoilers, automatic wheel braking and manual application of thrust reversers on the Pratt & Whitney engines quickly brought the aircraft to a halt, and I was able to turn off at approximately the halfway point on the 9,000-ft. runway.

Taxiing the aircraft was probably the most difficult aspect of the entire operation because the nose-wheel steering tiller was extremely stiff. I found that I could turn the aircraft much more smoothly by grasping the bar extending from the tiller hub rather than by holding the knob at the end of the bar. This was a small inconvenience.

Chapter 2

Medium commercial

MD-90 blends new, proven features

David M. North/Long Beach, Calif./Roswell, N.M.
October 17, 1994

Douglas was limited in the changes it could make without requiring a new type of rating for pilots, but managed to cut cockpit clutter while improving engines and systems.

McDonnell Douglas' MD-90 offers operators better performance and lower noise levels than its MD-80 predecessor, while retaining a common pilot type rating with the earlier DC-9 derivatives. The company's drive to retain a common type rating for the newest of the DC-9 series transports limited the changes that Douglas engineers could incorporate into the MD-90, especially in the cockpit. Airline pilots accustomed to flying the newer Boeing 757/767 and Airbus A320/340 transports will find the MD-90 cockpit a throwback to earlier aircraft that did not include large electronic displays and highly automated systems.

Pilots familiar with the DC-9 series, however, will be comfortable in the MD-90 cockpit and welcome the improvements. The Douglas designers, with the input of company and airline pilots, did improve visibility from the cockpit, reduce clutter on the instrument panel, and make sensible changes to the aircraft's system displays and controls.

The major changes in the MD-90 are not evident from the cockpit, but do add life to a line of aircraft that started with the DC-9-10 in 1965. The most significant change is the replacement of the 25,000-

McDonnell Douglas MD-90 includes a 4.75-ft. long fuselage plug and IAE V2500 engines. Other changes include a new APU, electrical power system and carbon heat sink braking system.

lb.-thrust Pratt & Whitney JT8D engines with International Aero Engines V2500 engines. The V2500 has a 25,000-lb.-thrust rating, with a 28,000-lb.-thrust rating optional. The new engines give the MD-90 noise levels well below Stage 3 requirements, and below potential Stage 4 levels. As an added plus, the V2500 engine also emits lower amounts of carbon monoxide, oxides of nitrogen and hydrocarbons.

Other improvements in the MD-90 include the installation of a new passenger interior and vacuum lavatories. A 4.75-ft. plug added to the standard MD-83 fuselage allows the MD-90 to fly with a mixed class seating of 158 passengers, or a maximum of 172 seats.

This *Aviation Week & Space Technology* pilot recently had the opportunity to fly the MD-90 from the company's testing site in Roswell, N.M. The day prior to the evaluation flight, I was briefed on the aircraft's systems and flew the MD-90 simulator at the Douglas plant at Long Beach. The fact that the two takeoffs and landings in the simulator were uneventful, and the landings were actually smooth, prepared me for the aircraft flight. I normally have more difficulty performing smooth landings in the simulator than in the actual aircraft.

The MD-90 flown on the evaluation was the second prototype, which flew for the first time one year ago. Following FAA certification, planned for later this year, this aircraft is scheduled to be deliv-

Interior of the MD-90, like other DC-9-series aircraft, has three seats on one side of the aisle and two on the other. New overhead racks and other improvements have been added in the cabin.

ered to Delta Air Lines. Douglas Moss, an experimental test pilot with McDonnell Douglas, was to fly from the right seat, while William Jones, the MD-90 chief experimental test pilot, was an observer. John Groves was the company flight test engineer.

The flight was the 556th for the aircraft, and it had accumulated close to 625 hr. prior to our flight. The first prototype MD-90 has logged more than 1,000 hr. in the flight test program. Gross weight at the blocks was 128,050 lb., which included 20,200 lb. of fuel. The second prototype is still equipped with flight test equipment in the cabin and ballast to increase the zero fuel weight. The gross weight was 82% of the maximum takeoff weight, while the fuel load was 52% of capacity.

A preflight walkaround of N902DC did not reveal any major changes from the MD-80 series aircraft, aside from the obviously larger IAE V2500 engine and nacelle, the new pylons and the less obvious 4.75-ft. plug in the midfuselage section. The additional length accentuates the comparatively small wings of the MD-90. They are essentially the same wings as those on the earlier DC-9 aircraft, with added strength. The wing size also partially explains the maximum 37,000-ft. certificated altitude for the MD-90.

Following the walkaround, I took the left seat and Moss the right. The pre-start checklist was accomplished quickly, while the Garrett GTCP 131-9 auxiliary power unit was powering the aircraft's electrical system. Both pilots said the APU had worked extremely well during flight test, and that it provided more generated power and had a lower specific fuel consumption than the APU in the MD-80. As an example of the reduction in cockpit clutter, the APU inlet door switch was removed from the overhead panel, and door operation is completely automated.

The left engine was started using the automatic sequence by pulling the start switch out and puffing the fuel switch on at 15% N_2. The exhaust gas temperature reached a maximum of 528C on the relatively hot day. During start of the right engine, we received an indication that the automatic abort systems had activated and that automatic clearing of the engine was taking place. A mechanic found the 10th stage bleed air valve was sticking open and corrected the fault by tapping it. The second right engine start was uneventful.

Moss entered the flight plan into the aircraft's dual flight management system through the multipurpose control display unit on his side. Douglas has made changes to the FMS by adding in-flight monitoring and an increased data base. At the same time, he programmed the FMS to automatically cut back the engine power ratio from a setting of 1.48 to 1.26 at 800 ft. The automatic power reduction feature can be used at noise sensitive airports.

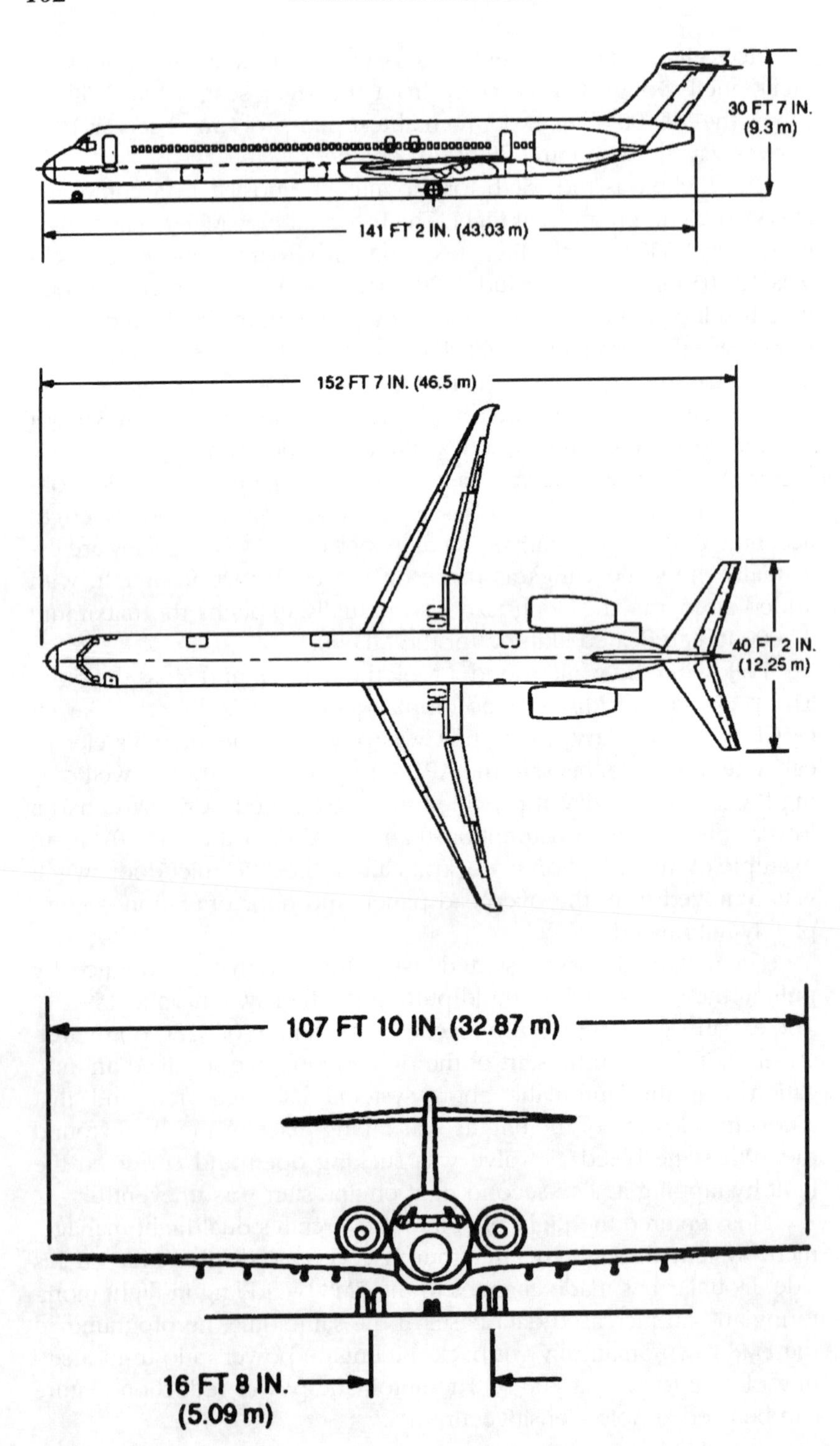

30 FT 7 IN.
(9.3 m)
141 FT 2 IN. (43.03 m)
152 FT 7 IN. (46.5 m)
40 FT 2 IN.
(12.25 m)
107 FT 10 IN. (32.87 m)
16 FT 8 IN.
(5.09 m)

I taxied the MD-90 to Runway 21 at Roswell. Aside from the occasional Air Force F-111 doing touch and goes at the airport, there was little traffic. The rudder pedals steered the aircraft more than adequately for most turns on the ground, and the tiller was only used for sharp turns on the ramp. The MD-90 is equipped with a new Aircraft Braking Systems Corp. carbon heat sink braking system that turned out to be very effective during taxi and on the final landing. Moss said the brakes had worked well during the maximum landing weight tests.

The temperature at Roswell at about noon was 92°F. The wind was from 190 deg. at 6 kt. The elevation of the former U.S. Air Force base is 3,670 ft. Graves calculated the V_1 decision speed to be 128 kt. and the rotation speed at 134 kt. The V_2 takeoff safety speed was 142 kt., and the calculated balanced field length was 6,050 ft.

As this was a Delta aircraft, the optional 28,000-lb.-thrust rating was chosen for takeoff by merely depressing the 28K button on the center instrument panel. The thrust rating is displayed above the engine instruments, and the appropriate power setting bugs are adjusted accordingly. The flap setting for takeoff was 15 deg.

I advanced the power levers to near the selected power setting and acceleration was brisk. A slow rotation to a 20-deg. nose-up attitude started 31 sec. after brake release, and initial rate of climb was 3,500 ft./min. after the landing gear and flaps were retracted. Moss noted the automatic cut back feature had deactivated during the takeoff roll, and he manually reduced power at 800 ft., lowering the climb rate to 2,500 ft./min.

The system worked as advertised on a later takeoff. There are multiple means of disarming the cutback system, including throttle movement after takeoff and sensor data loss. The system automatically disarms and advances power to the go around level if it detects an engine failure, wind shear or negative climb rate.

The MD-90 reached an altitude of 10,000 ft. 4 min. after takeoff. Air traffic control held us at that altitude for about 5 min. The autopilot and autothrottle were used for the climb out to 31,000 ft. The workload required to use the MD-90's automatic features appeared to be minimal as I observed Moss.

During the time at straight and level, I got a better look at the instrument panel. The Douglas designers removed the clutter of engine and system round dials from the center instrument panel of the MD-80 and installed electronic engine and systems display panels. The new engine displays with both digital and dial readings and the digital system readouts were easily assimilated during the cockpit scan.

Once again in a climb, at 15,000 ft. the rate of climb was 2,500 ft./min., and the fuel flow was 6,320 lb./hr./engine at an economical

The McDonnell Douglas MD-90-30 is the first of a new family of quiet and fuel-efficient twin-jet airliners powered by International Aero Engines (IAE) V2500 engines. The V2500 is the quietest engine in its class and meets U.S. and international Stage 3 noise level requirements by substantial margins. The first test aircraft, T-1, had its first flight on Feb. 22, 1993, beginning a thorough flight test program. Delivery of the 153-passenger MD-90-30 is planned to begin in 1994, and production is expected to continue into the next century. In addition, an extended-range MD-90-50 is being offered for delivery starting in 1995.

speed of 310 kt. A total of 2.5 min. were required to reach 20,000 ft. The fuel flow was 5,640 lb./hr./engine and the rate of climb had decreased to 2,200 h./min. Approximately 3,000 lb. of fuel had been used since engine start.

At 25,000 ft. the fuel flow was 4,970 lb./hr./engine with a speed of 307 kt. and a rate of climb of 1,500 ft./min. The cruising altitude of 31,000 ft. was reached 20.5 min. after takeoff, but that total included the 5 min. spent at 10,000 ft. Fuel flow just prior to level off was 4,370 lb./hr./engine, and the total fuel burn from engine start was 4,000 lb. The flight management system (FMS) indicated that 35,600 ft. would be the optimum altitude for our near 123,000-lb. gross weight.

The speed of Mach 0.75A used to reach 31,000 ft. was close to that identified by Moss as the long-range cruise speed for the MD-90. The outside temperature was –33C (–27F) or 13C (6F) above normal at the cruising altitude. Fuel flow at Mach 0.744 was 2,930 lb./hr./engine. The FMS indicated the maximum range speed was Mach 0.694. At that speed, the fuel flow was 2,550 lb./hr./engine. For a maximum speed cruise, Mach 0.793 was used. The fuel flow was 3,220 lb./hr./engine, and the true airspeed was 478 kt. Both Moss and Graves said fuel consumption figures for the MD-90 were generally 15% lower than those of the MD-80.

Another issue resolved in the MD-90 is the effect of fuel cold soaking on the wing. A return to tank fuel heating system has reduced cold soaking and eliminated the need for a separate fuel heating system.

At this point, Moss asked air traffic control for a return to Roswell. The MD-90 had precisely tracked along the flight path. The new transport is equipped with dual Honeywell Inertial Reference Systems as standard. The dual system was an option on MD-80s. The navigation system has been designed with provisions for Global Positioning System receivers to be added.

The MD-90's forward center of gravity required that a powered elevator be installed to win certification. The change was needed to improve pitch rate in a go around situation. Following some 60-deg. bank turns at 31,000 ft., Moss turned off the powered elevator. I found that elevator control was slightly more sluggish without power, but no more difficult to establish and hold. Moss said that at first the elevator control had been too quick, and had to be dampened.

The MD-90 also is equipped with power rudder control, with manual reversion. The rudder throw was increased to offset the higher thrust engines in an engine out situation. The ground and air minimum control speeds for the MD-90 are lower than those of the MD-80 because of the greater rudder authority.

While descending at Mach 0.74, deploying the speedbrakes did not cause any pitch movement and very little buffeting. Again at 15,000 ft., Moss turned off the powered elevator while I made some 60-deg. bank turns with the same result as at the higher altitude. The MD90's gross weight was now close to 119,500 lb. I retarded the throttle levers to achieve a 1 kt./sec. deceleration to enter a stall in the clean configuration in level flight. The MD-90's stick shaker activated at 160 kt. and the stall announcement came at 152 kt., with no tendency to fall off on a wing. Recovery was quick with a slight drop of the nose and steady application of power.

Moss lowered the landing gear and set flaps at 28-deg. to simulate an approach. The stick shaker activated at 116 kt., and the stall came at 106 kt. The wings rocked up to 5-deg. just prior to stall, indicating slight instability. At stall, I found that I was using almost full aft column to maintain attitude. Moss said the instability was more pronounced when entering the stall with induced yaw. The MD-90's stall tests confirmed that no stall warning computer changes were required and that the stall characteristics of the transport are equal to, or better than, the MD-80's.

While still at 15,000 ft., Moss retarded the left throttle, and there was more than adequate rudder trim while still in the dirty configuration to compensate for yaw.

During the descent into Roswell, Moss pointed out the changes in the MD-90's electrical system. The panel has been simplified. The MD-90 is equipped with a new engine-driven variable speed constant frequency power system that Moss said has proven to be very reliable while providing increased power. The new system also requires fewer pilot actions for normal and abnormal operations. The standard master electronic annunciator panel in the MD-90 is an improved version of the optional system found in the MD-80.

The first approach to Runway 21 at Roswell was in the autoland mode with a V_{ref} speed of 131 kt. The crosswind on approach was 12 kt., close to the 15 kt. maximum for selection of autoland. The approach was smooth, with the autopilot working to compensate for the bumpy air. The landing included a firm touchdown with the aircraft tracking down the centerline. The first three approaches were to a touch and go, but the autobrakes were not used in the final landing because the system needs more testing.

I made a manual approach and landing for the second touch and go. I found the aircraft to be responsive to power and control commands during the approach in turbulent air conditions. Earlier, Michael Wolf, an American Airlines MD-80 captain, suggested touching down with one main gear slightly before the other to attain a smoother landing in the MD-90. It appeared to work on my three landings.

The next approach was made with auto-throttle, which provided good pilot workload relief. The landing was again smooth. The final landing was made with the powered elevator off, and despite the requirement for slightly greater stick forces for control, it was similar to the other three. The brakes were effective in stopping the MD-90 without any tendency to grab at slower speeds. The new thrust reversers are an improvement over the earlier designs, Moss said. One change is that full reverse can be selected without waiting for the buckets to fully open.

Moss and Jones dropped me off at the takeoff end of the runway. They then took off almost over my head with the automatic cutback feature working to demonstrate the quietness of the MD-90. The aircraft's lock of noise was impressive.

Total time from engine start to shutdown was 2.7 hr. The actual flight time was 1.9 hr. The APU had been running for almost 1 hr. prior to our boarding and had been running during the flight. Three touch-and-go landings and two full-stop landings were made during the evaluation flight. Total fuel consumed was 15,000 lb.

The Douglas transports have a reputation for durability and long life, which explains why there are still more than 2,000 DC-9 and MD-

80 aircraft in service. The design service life for the DC-9 was 30,000 hr. and 40,000 landings, and for the MD-83 was 50,000 hr. and landings. The design service life for the MD-90 is for 90,000 hr. and 60,000 landings. The combination of durability, excellent performance, cabin flexibility and quietness inherent in the MD-90 should make it an excellent aircraft for operators.

While the internal requirement to maintain a common pilot type rating among the DC-9 series of aircraft has limited changes and improvements, the Douglas designers did what they could to ease pilot workload and improve situational awareness. The addition of the IAE V2500 engines is another improvement.

Transition training from other aircraft types is still under discussion with the FAA. The intent is that switching from the MD-88 to the MD-90 will take less than one day of training. The FAA pilots assigned to the MD-90 were proficient in 6 hr., Douglas officials said. Difference training from other MD-80 types would take five days of ground school and a proficiency check with two landings. Pilots new to the DC-9 series would take 102 hr. of ground training and simulator training.

MD-90 SPECIFICATION TABLE

McDonnell Douglas MD-90 transport is powered by two IAE V2525-D5 turbofan engines, each with a rating of 25,000 lb. of thrust. MD-90s to be delivered to Delta Air Lines will have V2528-D5 engines, each with a rating of 28,000 lb. of thrust.

WEIGHTS	
Maximum takeoff weight	156,00 lb. (70,760 kg.)
Maximum landing weight	142,000 lb. (64,410 kg.)
Maximum zero fuel weight	130,000 lb. (58,967 kg.)
Operational empty weight	88,171 lb. (39,994 kg.)
Fuel capacity	39,128 lb. (17,748 kg.)
Passenger capacity	158 mixed class
Cargo volume	1,300 cu. ft. (36.8 cu. meters)
DIMENSIONS	
Overall length	152.6 ft. (46.5 meters)
Wingspan	107.8 ft. (32.9 meters)
Overall height	30.6 ft. (9.33 meters)
Wing area	1,209 sq. ft. (112.3 sq. meters)
Sweep back at 25% chord	24.5 deg.
PERFORMANCE	
Cruise speed at 35,000 ft.	Mach 0.76
Range with full passenger and cargo/ domestic rules	2,170 naut. mi.

DC-9-80 nears Category 3 certification

Robert R. Ropelewski/Yuma, Ariz.
February 23, 1981

McDonnell Douglas Corp's Douglas Aircraft Co. subsidiary is nearing the design freeze point on the Category 3 automatic landing system and head-up display for its DC-9 Super 80 transport, which entered service last September (AW&ST Sept. 22, 1980, p. 30) certificated for Category 2 manual approaches and landings.

The Category 3 certification would allow DC-9 Super 80 crews to make automatic approaches and landings in weather with ceilings as low as 50 ft. and forward visibility of at least 700 ft. Category 2 minimums normally range from 100-150 ft., with forward visibility of 1,200-1,600 ft.; fully automatic landings are not permitted. Until the aircraft is cleared for Category 3 operations, the autoland mode of the DC-9 Super 80's Sperry digital automatic flight control system will remain inoperative on those aircraft being delivered to customer airlines.

This *Aviation Week & Space Technology* pilot recently flew the DC-9 Super 80 on a 2.5-hr. evaluation flight that included an examination of the aircraft's new digital flight control system, including the automatic landing function. Despite my limited preparation and briefings prior to the flight, I found the latest evolution of the aircraft's automatic flight controls and head-up display to be easy and comfortable to use, very flexible in terms of the autopilot capabilities available, and appealing in terms of the prospective workload reduction.

My evaluation flight began and ended at the Marine Corps air station here, where McDonnell Douglas operates a flight test facility for its new transport aircraft. I flew the No. 1 DC-9 Super 80 flight test aircraft-DC-9 airframe No. 909-with H.H. Knickerbocker, Douglas Aircraft Co.'s chief engineering test pilot.

The No. 1 Super 80 is the same aircraft that was involved in a landing accident at Edwards AFB, Calif., last May, when the tail of the aircraft separated from the rest of the fuselage after a hard landing (AW&ST May 12, 1980, p. 24). The hard landing was associated with Federal Aviation Administration landing distance certification testing. The aircraft subsequently was rebuilt and has been used since then as a test vehicle for the Super 80's digital automatic flight control system. The basic cockpit layout of the Super 80 is essentially the same as earlier versions of the DC-9, allowing pilots rated in those versions to transition to this aircraft without a separate type rating. Beyond the basic layout, however, the cockpit incorporates a considerable number of work-saving features, including:

- A thrust management system with full-time automatic throttles and a thrust rating computer that automatically computes required thrust and issues throttle commands for all phases of flight from takeoff to touchdown. An automatic reserve thrust capability is incorporated for engine-out situations, and an electronic engine synchronizer automatically eliminates engine "throb" in the cabin by matching engine pressure ratios (EPR) during takeoff and climb or go-around, or by matching engine rpm during all other flight conditions.
- Fuel control panel providing continuous digital fuel quantity readouts and continuous gross weight readouts based on the zero fuel weight of the aircraft as entered into the system's computer prior to takeoff.
- Dual overhead-mounted automatic pressurization system controls that require only destination airport altitude and barometric setting inputs (if available) prior to takeoff, with no further attention necessary during the flight. All new DC-9s are getting this system.
- Single test button for all caution and warning lights in the cockpit.
- Redesigned landing gear uplock latches that eliminate the need for the crew to conduct an up-latch check after takeoff. This check involves moving the landing gear handle to the up-latch position after the gears have been retracted on takeoff.
- Landing gear warning horn and lights inhibitor that suppresses the horn and lights when thrust is reduced significantly at airspeeds above 250 kt. This eliminates the need for the crew to cancel the horn manually under these circumstances.
- Automatic shutdown feature in the auxiliary power unit (APU) eliminates the need for the crew to monitor APU temperatures before shutting off the unit after the engines have been started and the electrical/hydraulic load has been shifted from the APU to the engines. A cooling off period of about 1 min. is needed for this on earlier versions of the DC-9. On the Super 80, the APU off position can be selected immediately after transferring the load to the engines, and the unit will turn itself off automatically after cooling down.

I took advantage of all of these features during the evaluation flight in the No. 1 DC-9 Super 80. The aircraft only recently came back on flight status after being grounded for several weeks for the installation of higher-thrust Pratt & Whitney JT8D-217 turbofan engines. Initial production versions of the Super 80 are equipped with JT8D-209 engines, which produce 18,500 lb. takeoff thrust each up to 77F. The JT8D-217 generates 20,000 lb. thrust for takeoff.

APU and engine start were simple and routine, and we began taxiing to Runway 3R at Yuma with the aircraft at a gross weight of approximately 119,900 lb. (54,500 kg.), including 39,270 lb. (17,850 kg.) of fuel. The DC-9 Super 80 with JT8D-217 engines has a maximum takeoff weight of 147,000 lb. (66,800 kg.).

Pre-taxi and pre-takeoff checklists were expedited by several other features in the Super 80, including:

- Thrust limit index arrows on the EPR gauges that were set automatically at the maximum EPR takeoff limit when the takeoff button on the thrust limit display was pushed. There was no need to compute and set the bugs manually. Climb, maximum continuous cruise and go around settings can be selected similarly on the thrust limit display, located on the center instrument panel.
- Hydraulic system selection is completely automatic, and there is no need to select left, right or both hydraulic systems as on earlier DC-9s.
- Mach trim compensator test required on earlier DC-9s has been eliminated because Mach trim is now monitored by computers in the digital flight guidance system. Likewise, the speed command test function and autopilot servo switch function and indicators found in earlier DC-9s have been eliminated because of automatic monitoring by the flight guidance system.

Flaps and slats were extended prior to taxiing, using the Super 80's dial-a-flap system. There is no longer a separate slat handle, and the slats are extended with flap handle actuation. In addition to the standard flap handle, the flap control box on the Super 80 also has a small dial that can be used to select any flap angle from 0 to 25 deg. for takeoff. This will allow crews to optimize flap settings for their particular airline's operations. Flap extensions beyond 25 deg. will still require use of the flap handle. We selected 15 deg. on the flat dial for takeoff, and in addition to extending the flaps, this also extended the leading edge slats to their fully open 21-deg. position.

As Knickerbocker completed the pre-takeoff checklist from the right seat, I taxied the aircraft onto the runway and pushed the takeoff/go-around palm switches at the rear of the throttles. As I put my hand back on the control wheel, the throttles advanced automatically to a pre-computed EPR limit of 1.93 and maintained that setting as the aircraft accelerated on the takeoff roll. Knickerbocker called V_R at 128 kt., and I pulled back on the wheel to establish a climb attitude of approximately 20 deg. nose-up after a takeoff roll of approximately 4,000 ft.

This particular aircraft was equipped with a Sperry Split-Cue Flight Director System featuring a 5-in. attitude director indicator and 5-in. horizontal situation display. Once the climb was established, I followed the command cues of the flight director.

Control forces are distinctly heavier on the Super 80 than on earlier versions of the DC-9, and the aircraft generally has the feel of a larger transport aircraft in all phases of flight. While this required some additional muscle during maneuvering at low speeds, it also translated into better stability characteristics during both the low- and high-altitude phases of the flight.

We turned northward after takeoff to begin the first leg of a flight to Palmdale Airport, about 250 naut. mi. away. After climbing 1,000 ft., Knickerbocker selected the climb mode on the thrust limit panel, and the throttles automatically reduced to the new EPR climb limit.

Under the control of Los Angeles Center, the aircraft climbed in steps to 28,000 ft., exercising the heading select, indicated airspeed hold and altitude hold functions of the Sperry autopilot along the way. Both airspeed and Mach number could be preset and selected using the same push/pull and turn knob on the left side of the glareshield-mounted flight guidance control panel. For the initial climb, an airspeed of 250 kt. was selected and then was maintained by the autothrottles. Climb rate was determined by the EPR limit in the thrust management computer.

The heading select portion of the flight guidance control panel included an inner knob permitting selection of bank angle limits from 10 to 30 deg. in increments of 5 deg., and an outer knob for selecting the desired heading. The outer knob had to be pushed in to engage heading hold, and this disarmed any other direction control mode, such as VOR tracking.

An altitude hold panel farther to the right on the flight controls panel allowed preselection of assigned cruise altitudes. As the aircraft approached those altitudes, it automatically began a smooth level-off, and the throttles backed off automatically to maintain the selected altitude and airspeed.

A relatively large flight guidance mode annunciator panel was located at the upper inboard section of the instrument panel on each side of the cockpit, and these were divided into four separate sections to provide continuous advisories of the mode and status of the autothrottle system autopilot/flight director, and roll and pitch axis controls. The heavy transport stability of the DC-9 Super 80 was evident during this portion of the flight. The aircraft was buffeted very little despite 60-70 kt. winds at our altitude and rugged terrain below us.

After stabilizing the aircraft in cruise at 28,000 ft. and approximately 0.78 Mach, and while I was flying the aircraft manually, Knickerbocker reduced the thrust to idle on the right engine, simulating an engine failure. There was negligible adverse yaw and little difference in performance except for the need to increase thrust slightly on the left engine to maintain speed and altitude. The right engine was then brought back to speed.

Arrival at Palmdale was essentially the reverse of departure from Yuma in terms of the use of the automatic flight control system. With the autopilot and autothrottles engaged, it was necessary only to dial in the desired airspeeds and the altitudes and headings assigned by Edwards Approach Control, which serves the Palmdale Airport.

Radar vectors took us to and through the inbound instrument landing system approach course to Runway 25 at Palmdale, but approach controllers finally steered us to a heading of 180 deg. to intercept the localizer and glideslope. At this point we were still descending to our initial approach altitude, using the vertical speed mode of the flight guidance system with a selected descent speed of about 2,500 ft./min. Heading hold was engaged to keep a 180-deg. heading, and the aircraft was slowing to 180 kt. using the speed select capability of the guidance system.

Both VHF navigation sets were tuned to the Palmdale ILS, the final approach course of 250 deg. was dialed into the course selector of each set, and the autoland mode selector was pushed. All other modes except go-around were inhibited from that point. The flight mode annunciator indicated that the automatic landing mode was armed, and the heading select and vertical speed modes were disengaged.

The ILS localizer and glideslope were captured almost simultaneously, and the flight mode annunciator indicated localizer and glideslope capture during the intercept process, and then localizer and glideslope tracking once the aircraft had stabilized on the ILS approach.

Winds at Palmdale were reported from 180-190 deg. at 14-17 kt., and this resulted in the aircraft crabbing to the left approximately 7-8 deg. At an altitude of 145 ft. on the radio altimeter, the autopilot began to transition the aircraft from a crab to a side slip, with the left wing down and right rudder applied to keep the aircraft aligned with the runway centerline despite the wind. This was accomplished automatically.

At 50 ft. above the runway, the flare sub-mode was engaged automatically and the throttles began to retard as the autopilot slowly raised the nose of the aircraft to slow its descent. When the aircraft touched down it was 10-15 ft. to the right of the runway centerline.

Throttles continued to retard to the idle position, and as the wheels spun up, lowering of the nose was initiated. In the basic Super 80 autoland system, the autopilot is programed to disengage 5 sec. after wheel spin-up. This aircraft was equipped with an optional rollout system, however, that provided rollout guidance down to nearly a full stop. All that was necessary was to apply reverse thrust with the engines and apply wheel braking. An automatic braking option is available, although it was not installed on this aircraft.

After coming to a stop and taxiing back to the downwind end of the runway, we took off for another approach. As in the first takeoff, the takeoff button was pushed on the automatic thrust computer, and the computer then set the reference indices on the EPR gauges. Depressing the takeoff/go-around palm switch on the back of one of the throttles activated the throttles, and they both advanced to the takeoff power setting.

As the aircraft lifted off the ground, however, Knickerbocker brought the right engine back to idle, simulating another engine failure. At our low speed of approximately 130 kt., there was a substantial yaw to the right that required some heavy pressure on the left rudder to counteract. Once the rudder was trimmed, this was not a problem.

At its relatively low weight, the aircraft climbed and accelerated well on the single engine, and we flew the aircraft in this condition until beginning the turn to intercept the ILS for another approach and landing. This time, the head-up displays were swung down from their overhead stowage, the autopilot was disengaged after localizer and glideslope capture, and I flew a manual approach and landing using the HUD cues for guidance.

The original purpose of the HUD in the DC-9 Super 80 was to serve as a monitor for the automatic landing system, but the display appears to be an effective tool on its own when a fully automatic landing would not be feasible or desirable.

The main elements of the display are a rectangular box representing the ILS window in which the aircraft had to stay, a small command dot serving the flight director function, and a small circle with wings representing the aircraft. Keeping the aircraft symbol over the command dot maintained the aircraft on its proper flight path for the approach. Allowing the command dot to drift away from the aircraft symbol generally resulted in both symbols departing the ILS window, but reimposing the aircraft symbol on the command dot brought both of them back into the window in a few seconds without too much difficulty. A decision height of 250 ft. was preset on the radio altimeter, and the letters DH began blinking at the right side of the bead-up display when we reached that altitude on the approach.

At 100 ft., the box depicting the ILS window disappeared and was replaced by symbolic runway sidelines providing a perspective display of runway alignment. As we descended below 60 ft., the command dot began flashing as a flare alert. As the altitude passed below 50 ft., the dot stopped flashing and began to rise on the display. Pulling back on the control wheel to keep the HUD's aircraft symbol over the command dot resulted in a smooth flare and touchdown. After lowering the nose, I depressed the takeoff/go-around palm switches on the back of the throttles to initiate an immediate takeoff.

The HUD provides takeoff and go-around guidance as well as landing guidance, and I used the same technique of keeping the aircraft symbol over the command dot to establish climb attitude after the touch-and-go landing. All of our approaches at Palmdale were flown at reference speeds plus gust correction factors that give us approach speeds of 125-130 kt.

A final approach to Palmdale was performed using the full capabilities of the autopilot and autoland system, with the HUD employed as a monitor. The approach and landing were textbook perfect, with a smooth, if not somewhat firm touchdown on centerline approximate 1,200 ft. down the runway.

Because of the ease with which crew members can set and forget desired flight parameters in the DC-9 Super 80, the digital flight control system also has some built-in protective devices. These are aimed primarily at avoiding serious underspeed or overspeed conditions. In particular, the system has an angle of attack floor that varies with flight conditions and the rate of change of angle of attack.

On the final approach to Palmdale, for example, the airspeed select was deliberately set at 100 kt. with the flaps at only 15 deg. Normal approach speed should have been 124-125 kt. with flaps at 40 deg. As the aircraft slowed to 135 kt., the throttles advanced automatically to hold that speed, which is about 1.4 times the stall speed of the aircraft in that configuration.

While cruising at 330 kt. at 19,000 ft. on the return flight to Yuma, I dialed 360 kt. into the speed select panel. This is 20 kt. above the normal maximum operating speed of the Super 80. The throttles advanced initially to accelerate the aircraft, but they retarded when the airspeed reached 340 kt., and the mode annunciator displayed a VMO LIM warning for maximum operating limit.

The Super 80 incorporates an elaborate stall warning system, and several stalls were executed on the return flight to observe the system. In an initial stall in the clean configuration at 19,000 ft., at an aircraft weight of approximately 105,000 lb., a distinctive buffet began around 151 kt. and the stall warning light and horn came on at 147 kt. Pushing the nose down promptly to a few degrees below the horizon resulted in a recovery with about 1,000 ft. loss of altitude. The aircraft slowed to 146 kt. before recovering.

With 15-deg. flaps, buffeting and stick shaker action began at about 116 kt., stall warning lights and horn came on at 108 kt., and a minimum speed of 104 kt. was reached before recovery began. A final stall in the landing configuration with flaps extended 40 deg. brought the stick shaker and buffeting at 105 kt., stall warning at 96 kt., and a minimum speed of 94 kt. before recovery.

Final landing at Yuma was made using the head-up display for a non-ILS approach after flying the latter portion of the flight on a single engine from the start of our descent from 19,000 ft. to the landing itself. There were no particular problems under these circumstances, although any power adjustments had to be accompanied by relatively large rudder inputs because of asymmetric thrust.

The automatic reserve thrust system increased the available thrust to 20,850 lb. on the good engine, and was more than an adequate thrust margin given the relatively light weight of the aircraft at that point.

Non-ILS approach symbology in the head-up display differs somewhat from that used for ILS approaches. Above 100-ft. altitude, there is no command dot or aircraft symbol, but only a horizontal line at the center of the display that defines the glideslope angle selected by the pilot. If the aircraft is flown so that the line is kept over the selected touchdown point on the runway, the aircraft will describe a path conforming to the slope selected by the pilot.

As in the ILS display, digital airspeed, attitude and vertical speed readouts are presented beneath the flight path symbology. At 100 ft. above touchdown, the flight path line changes to an aircraft symbol, and a flashing command dot appears in the aircraft symbol between 60 and 45 ft. At about 45 ft., the command dot stops flashing and begins to rise, providing a flare cue for the pilot to follow with the aircraft symbol.

The autothrottle was still engaged on the remaining operative engine for our final landing, and it was automatically retarded to idle as the aircraft descended below 47 ft. The overall result was a fairly precise approach and touchdown despite the absence of precision landing aids outside the aircraft.

Both McDonnell Douglas and FAA officials believe they are nearing certification of the bead-up display and the Category 3 automatic landing system on the Super 80, and FAA specialists said they expect a full-blown certification effort to begin in the next few weeks on this aspect of the DC-9 Super 80 program.

The company is still in the process of completing 500 automatic landings in a DC-9 Super 80 simulator to establish a data base from which the FAA will work. FAA flight test crews will then make about 100 automatic landings in the aircraft to verify that the actual landing results conform with those predicted by the simulator runs.

Nimble regional jet stable in all regimes

David Hughes/Wichita, Kan.
September 7, 1992

The Canadair Regional Jet 50-passenger transport is as easy to maneuver as a business jet yet provides stability in both high- and low-speed flight, which probably will appeal to airline operators.

The aircraft is designed to serve short routes into hubs that typically have been served by smaller turboprops. It also can serve city pairs separated by 1,200 or even 1,500 naut. mi. that are not large enough to merit service with larger transports.

This *Aviation Week & Space Technology* pilot evaluated flight test aircraft serial No. 7002 during a 2 hr. 30 min. flight from Wichita's MidContinent Airport. Doug Adkins, director of flight operations and chief test pilot for Canadair, and Roger Booth, senior flight test engineer, planned to show me both low- and high-speed handling qualities in a comprehensive look at what the aircraft can do.

The Bombardier Regional Aircraft Div.'s Regional Jet is powered by General Electric CF34-3A1 high-bypass turbofans with a ratio of 6.2:1. Each engine has 9,220-lb. of thrust.

Regional Jet No. 7002 is hangared at Bombardier's new flight test center at Learjet, where certification work has been underway for a year. Adkins said the Regional Jet has been one of the most trouble-free development projects he has been involved in, particularly from

Canadair Regional Jet, powered by General Electric CF34-3A1 engines, will have a range of 1,620 mi. and will cruise normally at Mach 0.74.

an aerodynamic standpoint. He added that the European Joint Airworthiness Authorities (JAA) evaluation team, which had just left Wichita, found no aerodynamic deficiencies to be corrected.

Transport Canada certified the Regional Jet July 31, and JAA certification is pending. The FAA sent a special team just to evaluate the electronic flight instrument system and is sending another team to evaluate the aircraft.

Werner Laatz, the flight chief for the Canadair Jet at Lufthansa CityLine GmbH., the European launch customer, stopped by Adkins office while I was there. Laatz said he was impressed with the simulator training course operated by Bombardier in Montreal using CAE Electronics simulators. He recently finished the course, including eight simulator sessions. CityLine pilots who are coming into the program out of turboprops will receive an additional three sessions to familiarize them with jet operations and the glass cockpit.

Laatz believes CityLine pilots experienced in turboprops equipped with electronic flight instrument systems will not have much problem adjusting to the Rockwell-Collins cockpit. CityLine captains transitioning into the Regional Jet have 5,000-10,000 hr. of flying time, while copilots have an average of 1,000 hr.

Laatz said the aircraft has delivered on its promise of being a fuel efficient transport. In fact, Canadair engineers were so conservative that the Regional Jet's fuel efficiency has turned out to be 10% better than anticipated under certain conditions. The aircraft is also "very stable in both high-speed and low-speed flight," Laatz added.

Adkins, Booth and I then headed out to No. 7002, which is still configured as a test aircraft with a flight test engineer station in back. Our ramp weight of 44,150-lb. was about 7,000-lb. less than a fully loaded RJ Series 100ER would weigh on the ramp. We had 8,000-lb. of fuel versus the full fuel load of 14,305-lb. (6,489 kg.). The ER version has a 4,998-lb. (2,267 kg.) capacity center wing tank in addition to the left and right wing tanks installed on the Series 100. Most customers are expected to order the 100ER version with the center fuel tank, which extends the aircraft's range from 980 naut. mi. to 1,620 naut. mi.

Our center of gravity was 21.8%, a fairly midpoint CG for an aircraft certified for flight from 7% to 35%. The CG could go farther aft without any control problems, according to Booth. The JAA team requested a 3-hr. flight with an aft CG in a lightly loaded RJ and found no problems, Booth added.

The basic Challenger airframe was stretched by adding a 128-in. plug forward of the wing and a 112-in. plug aft. Canadair engineers found they could achieve a 3-4% performance improvement just by

Lufthansa CityLine's Werner Laatz said he is impressed with the Regional Jet's performance and stability.

moving the engine about 9 ft. aft from its position near the wing. This provides for a better load distribution over the wing. During the test program it was discovered that the elevator would have to be increased in size as well, and its chord was extended by about 30%.

Adkins showed me the data traces of one stall event that came precariously close to a deep stall before the elevator was modified. A third power control unit (PCU) was also added to both the left and right elevators to handle the enlarged surfaces, which are driven now by all three hydraulic systems.

Each Regional Jet wing also has two ground spoiler panels, instead of the one found on the Challenger, as well as spoilerons and a flight spoiler panel which can be extended anywhere up to 40-deg.

Canadair engineers looked at every aspect of the aircraft to ruggedize it for commercial airline service. Instead of flying 300-600 hr. per year, as the average Challenger does, the Regional Jet will fly about 3,000 hr. per year. A new access door will allow mechanics to remove and replace avionics boxes without having to enter the aircraft cabin. The aircraft acrylic wind screen with a heating system for shedding rain also has been replaced by a more durable glass one with windshield wipers. The RJ cockpit is large enough to be comfortable, and the Rockwell-Collins Pro Line 4 glass cockpit has the look and feel of a large transport system.

Adkins performed a normal engine start by arming the igniter and activating the engine start switch, then advancing the thrust lever to idle about when the N_2 rotor rpm reached 20%. Starter shutoff occurs at 55% N_2. Following startup, fan vibration monitoring indexes for both engines appear on the left engine indication and crew alerting system (EICAS) display.

The two EICAS displays in the middle of the panel provide pilots with a wealth of data. The left, or primary, display shows engine indications and other data while the right "status" display presents synoptic diagrams showing the real-time status of electrical, hydraulic, fuel, flight controls and anti-ice systems. Adkins scrolled through a few of these to perform his checklists and used the display to check control surface position as he moved his wheel.

We received our taxi clearance and I advanced power slightly to begin rolling on the ramp. I used the steering tiller, which seemed sensitive at first but was easy to get used to. The aircraft is quite nim-

Collins Pro Line 4 glass cockpit has six displays, including primary flight display (inset left), primary engine indication and crew alerting system (inset center), and seconday EICAS that shows various synoptic diagrams (inset right).

ble during taxi, and it will be easy for airline pilots to maneuver in and out of crowded romp areas.

The primary flight display combined attitude and horizontal situation displays as well as speed and altitude tapes and a vertical speed indicator in the shape of a round dial. We planned a takeoff from Runway 14 with a temperature of 76°F. The winds were calm and the ceiling was 2,400 ft. and broken. Our speed settings were 125 kt. for takeoff decision (V_1); 129 kt. for rotation V_r and 136 kt. for takeoff safety speed (V_2). Flaps were set at 20 deg.

We taxied down Runway 32, reaching a speed of 60 kt. so I could evaluate the brakes, which caused a rapid but smooth deceleration. Near the threshold I executed a 180-deg. turn. The aircraft responded nicely, and the nose gear did not scrub.

Once we were cleared for takeoff, I released the brakes and began advancing the throttles to 91.1% N_1. I rotated the nose at 129 kt. and soon reached the 15-deg. nose up attitude recommended after takeoff. I maintained runway heading, and Adkins retracted the gear, followed at about 400 ft. agl. by the flaps.

The Regional Jet had power to spare at our broke release gross weight of 43,841 lb. Shortly after liftoff we were climbing at 4,500 ft. per min. as the engines each burned 2,900-lb. per hr. at 90% NJ. Passing through 8,000 ft. we were still climbing at 3,100 ft. per min.

We headed out to the Boot Hill test area near Wichita. At 15,000 ft. we were burning 1,979-lb. per hr. per engine and had consumed nearly 600-lb. since brake release. The temperature was 15°C above ISA standard conditions.

At 20,000 ft. we engaged the autopilot, which handled the rest of the climb smoothly. The Collins two-axis APS-4000 autopilot has been certified by Transport Canada, but additional work is needed to smooth out some of the system's performance and complete all Category 2ILS requirements.

We were climbing at 800-1,000 ft. per min. passing FL 300 and burning 1,560-lb. per hr. per engine. At 22 min. into the climb we passed FL330, maintaining Mach 0.69. We reached our final cruising altitude of FL370 29 min. after brake release having burned about 1,700 lb. of fuel with an average condition of ISA plus 12-13C. The engines were burning 1,160-lb. per hr. each, and we were climbing at 500 ft. per min. right before level off.

We had experienced two intermediate level offs on the way up and I had not adhered to the recommended speed schedule at all times, so the climb performance was not representative of the PJ's potential. However, the climb was not atypical of what might be encountered in airline service.

We pushed the power up to maximum cruise at 93.5% N_1 and accelerated slowly as the Regional Jet sought its equilibrium speed. We were established in a block altitude from FL350 to FL370, and I performed a 30-deg. bank left turn while Adkins pointed out that the aircraft does not lose much energy in a turn at this high altitude. "It's a good wing for that," he said. In fact we continued to accelerate slowly with the power up during the 180-deg. turn.

It took us 9 min. to accelerate out to Mach 0.774 at 450 kt. true airspeed. Later on the ground, Adkins looked up in a computerized performance table to find a climb based on similar temperatures. He found that the Regional Jet should take 34.6 min. to reach FL370 and then accelerate out to Mach 0.74 while covering 235 mi. all told, which was close to what we experienced.

The aircraft is capable of rapid roll rates at altitude, and there is a strong dihedral effect involving uncommanded roll in the direction of sideslip. Adkins demonstrated this by stepping on the rudder pedal. The aircraft began rolling off in the direction of the sideslip. He said the rate can reach 20-30 deg. per sec., which is typical of a swept-wing aircraft with dihedral.

I also explored the buffet boundary at high altitude by making several 60-deg. bank turns and pulling about 2g to engage the stick shaker. The stick shaker is programmed relative to angle of attack to precede the buffet at high altitude. This alerts the pilot so he can keep the aircraft from penetrating the buffet region. By pulling a little more on the control wheel, the buffet could also be heard and felt.

After this we leveled the wings and executed a descent using spoilers which do not create much vibration even when activated rapidly. I tried putting them out rapidly and then slowly, as might be done for passenger comfort. The average increase in descent rate with spoilers is a healthy 1,600 to 1,800 ft. per min. Spoiler activation caused only a slight nose down pitch.

Next we examined what might happen if a pilot inadvertently left the power back while engaging the autopilot in vertical speed climb at 500 ft. per min. As we climbed under autopilot control with the throttles at a low setting, our speed bled off until a green bar appeared, which signified 1.3 stall speed (V_S). At 8-deg. angle of attack, continuous ignition engaged automatically as a precaution against flameout. At 9-deg. angle of attack the stick shaker activated. It is set at 1.07 V_S. At this point a warning beep sounded as the autopilot automatically disconnected and the aircraft pitched over until it was slightly nose down.

We recovered, and Adkins pointed out that the autopilot is designed not to stall the aircraft in a situation like this where the pilot forgets to add power. The autopilot only provides pitch trim at the

Fifty-passenger Canadair Regional Jet will be able to take off from 5,265-ft.-long runway at maximum gross weight and reach 35,000 ft. in 21 min.

rate of 1-deg. per sec., which did not keep up with our rate of deceleration. This is why we pitched over a bit even though the stick pusher did not engage.

Adkins said the British Civil Aviation Authority (CAA) was concerned about the possibility of the aircraft encountering negative g forces when the pusher activates at high altitude. To avoid this problem, the pusher on the Regional Jet has been programmed to pitch the aircraft forward but then to disconnect it before nose down moment results in negative g.

To set up the next demonstration we asked Kansas City Center for permission for a simulated emergency descent, which was approved. We pushed the nose over at Flight Level 370 to pick up a high speed descent. Our Mach number soon climbed post maximum operating speed (MMO), which was Mach 0.85 at our cruising altitude. We continued to accelerate past Mach 0.85 in a 25-deg. nosedown attitude until the vertical velocity indicator pegged out at 6,000 ft. per min. Our exact rate of descent could not be determined beyond that point.

The noise level in the cockpit picked up considerably and we then deployed the spoilers, which caused only a slight nosedown pitch. The deceleration was negligible as the panels did not deploy all the way due to air loads. At MMO they would deploy fully.

We ultimately reached Mach 0.91 in this rapid descent without losing any pitch or roll effectiveness in the controls before pulling out and leveling off at 16,000 ft. Adkins said that during the test program he has been as high as Mach 0.98. The Regional Jet is even more stable in the high Mach regime than the Challenger, and Adkins believes its high-speed handling characteristics are outstanding.

At 15,000 to 16,000 ft. I then flew a series of power-off stalls at a 1 kt. per sec. rate of deceleration, starting in a clean configuration. I applied back pressure steadily and continuous ignition engaged prior to stick shaker onset at 130 kt. at 13.8-deg. angle of attack. A bit more back pressure brought on the pusher action at 123 kt. and 16.5-deg. angle of attack. The pusher caused a substantial nosedown pitch to 10-15-deg. below the horizon, and I attempted to recover the attitude too quickly and encountered the shaker once again. Adkins said the long stroke of the pusher in the Regional Jet does a good job of putting the nose down, but the aircraft needs to accelerate before recovery.

With 20-deg. of flap the shaker was encountered at 114 kt. and 13.2-deg. angle of attack while the pusher activated at 109 kt. and 15.8-deg. angle of attack. The final stall was entered with 45-deg. of flaps and gear down, during which the shaker activated at 108 kt. and the pusher at 101 kt.

Next I flew the aircraft as Adkins activated roll disconnect so that my control wheel only moved the left aileron. When I turned the wheel right, the right spoileron came up as well. This mode of operation is used to counter a jammed aileron. The aircraft was completely controllable in roll with only one aileron, but the roll rate was substantially reduced. However, Adkins pointed out that by using 60-deg. of wheel travel instead of 30-deg., a pilot could match the roll rate with two ailerons at 30 deg. of wheel travel.

By pressing a Roll Select switch on the panel in front of me, I was able to obtain authority with my control wheel over the left spoileron

Interior of the Regional Jet has a ceiling height of 6 ft., and the reclining seats have an ample, 31-in. pitch.

as well as the right, and the roll rate seemed to return to normal. Adkins pointed out that there are large deflections of the spoilers as aileron movement increases. "I'm a strong believer that spoilerons are a good design feature in aircraft," Adkins said.

Once the ailerons were reconnected, Adkins activated pitch disconnect, which is used to deal with a jammed elevator circuit so that my control wheel only activated the left-hand elevator. The pitch rate seemed to be substantially reduced, but more than adequate to fly the aircraft.

We then examined the aircraft's Dutch roll characteristics in both a clean and a landing configuration by turning off both yaw dampers. Once initiated, the roll was mild in a clean configuration and did not develop much. The landing configuration Dutch roll, on the other hand, quickly diverged from neutral and we ended up with 25-deg. bank angles in 3-4 sec. by letting it go. Of course an alert pilot could arrest the motion quickly, as Adkins demonstrated by zeroing out the roll almost instantly with a deft control wheel input. Adkins said there is nothing in the Challenger or Regional Jet Dutch roll which poses a pilot induced oscillation hazard. The result is the Regional Jet can be dispatched with one yaw damper inoperative because it can be flown, hands-on, if the remaining one should fail inflight.

Following this exercise we contacted ATC and canceled our IFR clearance with Kansas City Center for a VFR descent into Salina Municipal Airport. I established the aircraft on a base leg to final to Runway 35, which is 13,337 ft. long.

The wind was at 280-deg. at 10-12 kt. and I flew this approach with 45-deg. of flaps with a reference speed of 129 kt. for the approach. As we crossed the threshold I had to fight the inclination to flare too high, based on a landing picture ingrained years ago when I flew the C-5. Even so, the touchdown was a smooth one and the aircraft proved easy to control at high speed on the runway. Our landing weight was 39,200 lb. at this point. Adkins reset the flaps and trim, and then I advanced the throttle for the go around.

The aircraft lifted off effortlessly and I was soon turning to begin another box pattern. The Regional Jet obviously had performance to spare at this weight, which prompted me to try quick, 45-deg. bank turns not unlike ones I flew years ago as a student in the T-38. The aircraft flew more like a jet trainer in the pattern than a large transport, showing there is plenty of maneuverability should an airline captain ever need to call on it.

I made the second visual approach and landing with a 20-deg. flap setting, which moved the nose attitude up quite a bit on approach.

During roll out I applied the brakes, which took hold quickly, and selected thrust reverse.

When you raise the reverse thrust levers there is a 2.5-sec. delay while the cowl moves back and the blocker doors go into place before the engines can go above idle. This was not a full stop, so I was soon stowing the reverse thrust levers and advancing the throttles for takeoff.

I maneuvered for a third landing at Salina using 45-deg. of flaps. This one would be a full stop landing followed by an immediate takeoff starting from the middle of the runway. Shortly after touchdown I hit the brakes hard. The stopping distance from that point was surprisingly short probably under 1,000 ft. Adkins said during the test program the Regional Jet often was able to make a turnoff to the Lear facility at the Wichita airport within 1,800 ft. of the threshold on Runway 19R. The stopping distance was short enough to draw favorable comments from airline pilots holding short for the same runway.

The heavy-duty Aircraft Braking System Corp. steel brakes are slightly over designed for the weight of the aircraft due to the short stage lengths and rapid turnarounds. Adkins said, "The intent was to have the capability of doing a fairly hard brake landing, taxi in and taxi out 15-20 min. later and do a takeoff."

After stopping on the runway, I began another takeoff by advancing the power. Adkins had briefed that he would simulate an engine failure at V_1. He retarded one of the throttles at about 112 kt. I rotated the aircraft gently and kept the nose-up attitude at about 5-deg. initially, although Adkins said I could have rotated as high as 10-deg. for an optimum engine-out climb.

Directional control was not a problem, as the two engines are mounted on the fuselage close to the aircraft centerline and the additional fuselage length of the Regional Jet gives the vertical stabilizer a lot of authority. He said some Challenger pilots who have flown the Regional Jet have even told him they prefer the larger transports handling characteristics.

We climbed out for a VFR return leg to Wichita, and Adkins and Booth demonstrated an engine shutdown and relight. We had a stable start, and the whole procedure took no longer than an engine start on the ground.

I made a straight-in approach to Runway 14 in Wichita and completed the flight with a uneventful full-stop landing. Again the brakes were very effective in stopping the aircraft quickly, and I taxied back to the Learjet facility. At engine shutdown the fuel remaining was 1,810-lb.

Bombardier's Regional Aircraft Div. has 36 orders and 36 options for the Regional Jet so far with Lufthansa CityLine in Germany accounting for 13 orders and 12 options. Comair, Inc., of Cincinnati, Ohio, has ordered 20 and placed options for 20. Comair is a Delta Air Lines affiliate with hubs in Cincinnati and Orlando, Fla.

CANADAIR REGIONAL JET SPECIFICATIONS

GENERAL

Twin-engine turbofan regional airline transport with a 50 passenger capacity. Certified by Transport Canada to FAR 25 standards. FAA & JAA certification pending. Approved for day/night VFR/IFR operations and flight into known icing conditions. Rockwell-Collins glass cockpit for a two pilot crew.
Base price (non-fleet)—$16 million U.S.

POWERPLANTS

Two General Electric CF34-3A1 high bypass ration turbofans with a 6.2:1 ratio. Static thrust at sea level with automatic power reserve (APR) 9,220 lbs. (4,182 kg.). Cascade-type fan air thrust reversers.

WEIGHTS

RJ Series 100ER (with additional center fuel tank)**

Typical operating weight empty	30,122 lb. (13,663 kg.)
Maximum zero fuel weight	44,000 lb. (19,958 kg.)
Maximum ramp weight	51,250 lb. (23,247 kg.)
Maximum takeoff weight	51,000 lb. (23,133 kg.)
Maximum landing weight	47,000 lb. (21,319 kg.)
Maximum payload	13,878 lb. (6,295 kg.)
Payload—full fuel	6,823 lb. (3,095 kg.)

**Weights not listed for the RJ Series 100 which does not have center fuel tank which carries 4,998 lbs. (2,267 kg.) of additional fuel. The RJ Series 100ER has this extra tank.

DIMENSIONS

Length	87 ft. 10 in. (26.77 m.)
Height	20 ft. 5 in. (6.22 m.)
Wing span	69 ft. 7 in. (21.21 m.)
Wing area (net)	520.4 sq. ft. (48.35 sq. m.)
Wing aspect ratio	8.85
Wing sweepback (25% chordline)	24.8 deg.

CAPACITIES

RJ 100ER Fuel	14,518 lbs. (6,585 kg.)
RJ100 Fuel	9,520 lbs. (4,318 kg.)
Baggage Total Gross Volume (Approx.)	503 cubic ft. (14.22 cu. m.)

PERFORMANCE

RJ 100ER Range, normal cruise	1,620 naut. mi.
RJ 100 Range, normal cruise	980 naut. mi.
High speed cruise	Mach 0.80
Normal cruise	Mach 0.74
Maximum operating altitude	41,000 ft. (12,496 m.)
FAR 25 Take-Off Field Length (MTOW)	5,265 ft. (1,605 m.)
FAR 25 Landing Field Length (MLW)	4,725 ft. (1,440 m.)
Fuel consumption at average cruise (at 6.7 lb./U.S. gal.)	525 U.S. gal./hr.

CERTIFIED NOISE LEVELS

Take-off	78.6 EPNdb
Approach	92.1 EPNdb
Sideline	82.2 EPNdb

Fokker 70 meets short-haul needs

David Hughes/Amsterdam
March 21, 1994

Fokker 70's speed protection system and effective speed brakes are two features well-matched to short-haul operations into busy hubs.

The Fokker 70, the Fokker 100's derivative, is well-suited to operate in busy terminal areas where rapid changes in ATC instructions and landing at ILS minimums are routine.

Fokker plans to file for a common type rating approval for the new 79-seat Fokker 70 and its cousin, the 107-seat Fokker 100. It would be difficult for an F100 pilot to spot any appreciable difference in the flying qualities of an F70.

The prototype F70 aircraft was flown recently by this *Aviation Week & Space Technology* pilot. It is the No. 2 Fokker 100 prototype modified by cutting out two fuselage sections. Wim J. Huson, an experimental test pilot with Fokker and head of flight test operations, said we would accomplish three test cards involving coupled ILS approaches during our sortie.

The stall strip on the wing has been increased in length by 50% to correct a tendency of the F70 to roll off a little more in the stall than the F100. Other than that, the F70 wing is identical to the F100.

Huson had requested that 16,094 lb. (7,300 kg.) of fuel be loaded on the prototype aircraft to achieve a ramp weight of 83,848 lb. (38,033 kg.). He said this would simulate a F70 full of passengers with enough fuel for two 1-hr. flight legs plus enough fuel to reach an alternate 200 mi. away with 45 min. of holding fuel. This prototype air-

Fokker 70 prototype, a modified Fokker 100, takes off from Amsterdam's Schiphol Airport as part of the flight test program.

craft is heavier than a production F70 due to the addition of flight test instruments and a water ballast system. Sjoerd Postma, a flight test engineer, also accompanied us on the flight.

I occupied the left seat and Huson the right as he showed me the cockpit that is indistinguishable from that of the F100. Fokker engineers have held strictly to the "dark cockpit" design philosophy. Some noncritical messages are inhibited when the aircraft is in critical phases of flight, such as when the aircraft is below 400 ft.

The six Collins EFIS displays (6-×-7-in. cathode ray tubes) on the front panel provided a wealth of flight and systems data. Huson keyed in our speed targets into the Honeywell flight management system (FMS), which is optional for the F70. The speeds included a V_1 of 125 kt., and a V_2 of 131 kt. If a pilot inadvertently selects a V_2 below stick shaker speed, it would be disregarded. A checking algorithm in the aircraft's automatic flight control and augmentation system (AFCAS) computer would rely on angle-of-attack guidance for the departure, ignoring an improper setting.

The F70 we were flying was equipped with an FMS with lateral navigation capability only, so Huson also set in the green dot speed for best performance, or 178 kt. The F100 FMS with VNAV mode does this automatically.

The engine start sequence was next. Huson started No. 2 engine and I started No. 1. I held the engine start selector switch momentarily to the No. 1 engine position to activate the starter motor. At the first indication of N_1 acceleration, I moved the fuel level to the open position. The turbine gas temperature peaked at about 440C.

After reviewing a standard instrument departure (SID) from Schiphol Airport's Runway 24, we were ready to taxi. As I initiated a 180-deg. turn, the tiller steering worked smoothly and braking seemed normal, even though Fokker has developed a taxi-braking function. Light braking causes wear on carbon brakes, so the taxi-breaking feature activates either the inner or outer brakes below about 20 kt. ground speed, but not both. This reduces wear.

As we taxied, ATC changed our departure to a SID from Runway 22. Huson armed autothrottles and the navigation function on the FMS. Once on the runway, our clearance was changed, with ATC telling us to fly direct to Spijkerboor VOR.

I stood the throttles up, let the engines accelerate and then pulled the takeoff/go-around (TOGA) triggers on each lever to engage the autothrottle system. The throttles declutched at 80 kt., and Huson called V_1 and then rotation. After I pulled up the nose and we achieved a positive rate of climb, Huson retracted the gear, and I followed the flight director command to 18-deg. of pitch. Huson engaged the autopilot shortly after takeoff and activated the heading mode to initiate a left turn as we passed through 500 ft. We were soon headed direct to Spijkerboor with flaps ups and autopilot engaged in the flight level change mode.

Passing the 10,000-ft. level, the fuel consumption rate was down to about 4,167 lb./hr./engine (1,890 kg.) as we accelerated toward 250 kt. As ATC gave us new headings and altitudes, it was clear the pilot could easily incorporate these changes on the glare shield-mounted flight mode panel and avoid going head-down to enter changes in the FMS control/display unit.

We were averaging 3,791 lb./hr./engine (1,720 kg.) passing 15,000 ft., and 3,120 lb. (1,415 kg.) passing 25,000 ft. at Mach 0.65. When we reached the top of climb at Flight Level 290, we had burned 2,398 lb. of fuel (1,088 kg.). After accelerating to Mach 0.75, we were burning an average of 2,270 lb./hr./engine (1,030 kg.), and we had burned a total of 2,546 lb. (1,155 kg.) of fuel since taxi start 29 min. earlier.

We then headed out over the North Sea to a Fokker test area. Huson had set up the Collins EFIS map display to show me the boundaries of danger area 6, which is used by the military for live fire exercises. We checked with ATC to make sure it was inactive before flying through. At the 29,000-ft. level we performed some steep turns. Huson performed a dynamic turn at the M_{MO}, of Mach 0.77 with a 2g pull to demonstrate how the wing provides good maneuvering capability for any emergency, such as collision avoidance. We did not feel any buffet. We were then at Mach 0.62, and I performed a 1.79 steep turn.

We then initiated a descent to the 10,000-ft. level, and Huson programmed in an 8,000-ft.-per-min. rate of descent—which he knew was beyond the aircraft's capability—to demonstrate the envelope protection features of the AFCAS. The aircraft began to accelerate, but we received a speed warning. The autothrottle system engaged automatically to reduce thrust to idle, and the AFCAS reverted to a flight level change mode, which targets a Mach speed with elevator changes. The result was we did not exceed the M_{MO} of Mach 0.77.

Huson then deployed the speed brakes, and the deceleration was smooth with hardly any buffeting. The clamshell-type speed brakes on the tail are large and effective and in an area where air flow is slower than over the wing. We reached a 7,000-ft.-per-min. rate, and we saw what an emergency descent might look like.

AFCAS also provides protection against flying too slow. To observe this, we started in level flight in a clean configuration with power idle and the autothrottles off. Once we slowed to reach the amber bar on the speed tape designating the minimum allowable airspeed (1.3 V_s in this case), the autothrottles engaged automatically and selected TOGA power. Thus we did not slow to the stick-shaker speed of 122 kt. (1.07 V_s) or experience any buffet, and I rotated the nose to an attitude of 25 deg.

While some larger transports do have more comprehensive envelope protection, the F70 system provides substantial capability for an aircraft in the 70-seat category (AW&ST Oct. 4, 1993, p. 40). Huson noted that the F70 speed protection is designed with three levels, including protection against inadvertent selection, warnings and finally automatic takeover or mode change when required.

Following this slow-speed sequence I started a descent toward Valkenburg airfield, a Dutch Royal Navy facility on the coast 15 naut. mi. southwest of Schiphol. While en route, I flew a simulated go-around to prepare for the next test. When I resumed the descent, the speed brakes came in handy as we were asked to expedite the approach. The F70 can decelerate in a descent. Huson showed me that if I forgot to put the speed brakes back in on the level-off they would close automatically. I could see the air traffic from London lined up

F70's clamshell speedbrakes, like those on the F100, allow deceleration in a descent, even in the terminal area.

going into Schiphol on my EFIS display with reports from the TCAS-2 system.

We planned to fly three ILS approaches to Valkenburg's Runway 23 as part of the F70 flight test program. The F70 is designed to help the two-man crew deal with short-haul flying that may typically involve six landings a day. In Europe where fog is prevalent at certain times of the year, 3-4 of these landings may have to be made at minimums. Due to Schiphol's proximity, the approaches had to be abbreviated. Our first approach would be with full flaps. We would be flying V_{REF} plus 5 (or 5 kt. above the top of the amber strip on the speed tape) with a weight of 79,806 lb. (36,200 kg.).

We were on radar vectors, and Huson selected flaps 8-deg. and suggested I use the speed brakes. It is possible to use speed brakes on a Category 2 autoland approach in the F70, giving pilots great flexibility in expediting descents in the terminal area. The aircraft can bleed off a lot of speed yet still achieve precision tracking for autoland.

Clouds were broken at 1,100 ft. and winds were 200 deg. at 10 kt. gusting to 19 kt. as I intercepted the localizer and glideslope. I was stabilized on final approach with flaps at 42 deg., following the flight director with autothrottles on for the test, and when we reached 150 ft. Huson called minimums and I activated the TOGA triggers. The throttles responded promptly, and I followed the flight director to a go-around attitude as Huson raised the gear and flaps. The attitude commanded a 2,000-ft.-per-min. vertical velocity designed to keep level-offs smooth rather than abrupt. At this rate of climb the F70 accelerates.

On the next approach we configured with flaps 25-deg. and gear down with one autopilot engaged for a coupled approach to 50 ft., flying V_{REF} plus 20 kt., or about 146 kt. Below 1,000 ft. on the approach both autopilots engaged to provide redundancy, and the autothrottle system slowed the aircraft to V_{REF} plus 10 kt. automatically below 500 ft.

The go-around was to occur at 50 ft., but I did not activate the TOGA triggers until 40 ft. The automatic power application and rotation of the nose to the go-around attitude was impressive because it was both prompt and smooth. Huson said we dropped another 15 ft., to 25 ft., above the runway before climbing. He said if TOGA triggers were activated at 15 ft. above the runway, the aircraft would only settle 8 ft. before climbing. On the go-around Huson raised the gear and flaps.

As we set up for the next approach, we configured early to 25-deg. flaps and gear down, and picked up V_{REF} plus 5 kt., or 128 kt., for the autoland approach to a full stop. Autoland is optional on the F70. We had been underway for one-and-a-half hours including taxi time at Schiphol, and we had burned 6,172 lb. (2,800 kg.) of fuel. Huson noted that the F70 can hold a higher airspeed as instructed by ATC while setting up for an autoland approach, as long as it crosses 800 ft. above the ground at 160 kt. or less. This would allow the aircraft to carry 180 kt. or more to the outer marker and slow to approach speed in time to complete the autoland.

The aircraft stayed well aligned on the glideslope and localizer throughout the approach despite the gusty wind, and we touched down and tracked within a meter of the centerline. With the thrust reversers activated, I stayed off the brakes and we continued to roll out down the 8,000-ft. runway disconnecting the autopilot shortly before turning off the runway.

We taxied back for another takeoff and set up to fly two visual patterns. After beginning my first turn to final I realized I was angling in and corrected to the left. Additional glide path corrections were needed once I was established on short final with full flaps. I did not flare as much as I should have, and so the touchdown was firm rather than smooth.

On the next pattern I made a better turn to final and was able to stabilize on a visual glide path with VASI guidance. At about 125 kt. I worked to adjust for the gusty winds, and the flare worked out better as far as rate of descent. After touchdown the thrust reversers were used to slow us down. We had used 7,581 lb. (3,439 kg.) of fuel since leaving the ramp at Schiphol 2 hr. earlier.

After landing I vacated the left seat to allow another pilot to fly the leg back to Schiphol, as previously planned. At shutdown we had been operating the aircraft for 2 hr. 30 min.

Joint sales of F70, F100 look promising to Fokker

David Hughes/Amsterdam

The 70-79 seat Fokker 70 could play a key role in the Dutch aircraft manufacturer's recovery if the company's analysis proves correct that this is the right size transport arriving on the market at just the right time.

The new Fokker 70, now being sold in joint sales with the Fokker 100, is expected to account for one-third of the jet transports (F70s and F100s) the Dutch aircraft manufacturer will sell over the next five years. By adding a smaller-size aircraft, Fokker has created a family of jet transports to broaden the appeal of the Fokker 100, which has accounted for about 300 sales over the past six years. Fokker officials said some customers have recently became interested in the F100 pri-

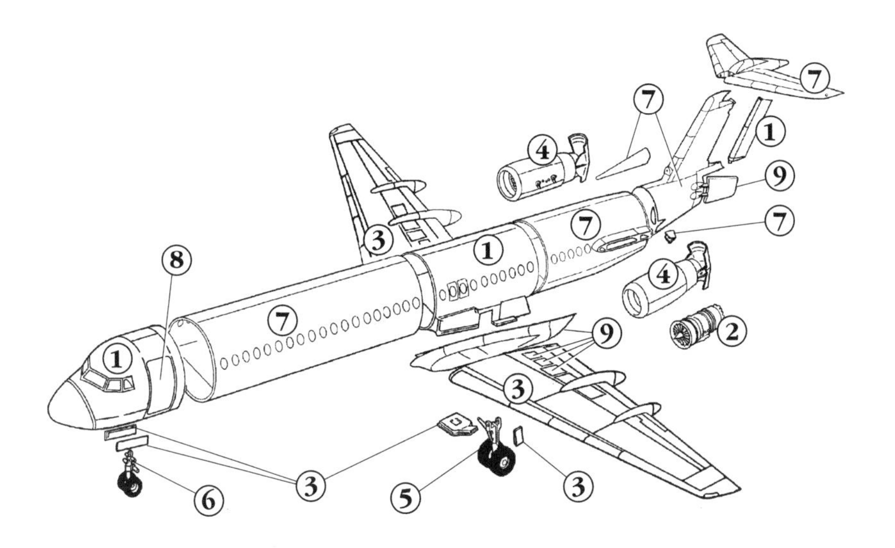

1. Fokker	4. Grumman	7. Deutsche Aerospace Airbus
2. Rolls-Royce	5. Menasco	8. Singapore Aerospace
3. Shorts	6. Dowty Aerospace	9. IPTN

Workshare for the Fokker 70 will match that of the Fokker 100, as both have 85% of parts in common.

marily because it can now be purchased in combination with the smaller F70.

Fokker has been going through difficult times after accumulating as many as 32 unsold commercial transports in early 1993. The company still has six inventory aircraft and is leasing nine F50s and six F100s to airlines (AW&ST March 14, 1994, p. 39). Fokker reduced employment by 2,000 last year and expects to cut 1,900 more this year, to reach a level of about 8,500.

The Fokker 70 is a derivative aircraft project, which cost approximately $180 million, and Fokker would not have brought it to market quickly if officials did not believe demand was strong. The project was announced at the Paris air show just last year, and an F100 was modified to match F70 dimensions so the flight test department could gain a six-month head start on the flying program.

J. Mel Kroon, vice president of marketing for Fokker, said interest remains strong in the aircraft, with 22 orders from airlines last year and 5 more in the first two months of this year.

The F70 is not without competition. Avro International, a British Aircraft Ltd. Div., is also offering four-engine jet transports, the RJ70 and RJ85, with 70 and 85 seats respectively. In the 50-set jet category, there is the Canadair Regional Jet with 29 aircraft in service and another 40 on order. Embraer is also developing a 50-seat jet as well, the EMB-145, for delivery in 1996 (AW&ST June 21, 1993, p. 28).

Whether the 70-seat market niche turns out to be the right one for Fokker is yet to be seen. Bombardier Aerospace is wrestling this year with the issue of whether it should launch a 70-seat stretch version of its Regional Jet, but Bombardier officials have said they do not see this niche gaining strength until later in the decade. A Fokker official said the Dutch company will also decide this year whether to develop the Fokker 130, a stretched F100 with 130 seats.

Having a 70-seat companion aircraft to the F100 gives Fokker flexibility in its marketing and production at a time when airlines are responding to rapidly changing market conditions. Airlines that order Fokker 70s and 100s can take unspecified options and then decide at a later date whether to take delivery of the F70 or the F100. The F70 costs $23 million and the F100, $33 million.

Fokker 70 cockpit is nearly identical to the Fokker 100 (left), based on a core avionics suite that can be upgraded.

Fokker expects the market for commercial transports and regional airline aircraft to begin recovering over the next two years. Production rates for the F70, F100 and the turboprop-powered F50 are set at a combined rate of 40 aircraft in 1994 and 1995, with the mix being about one-third turboprop aircraft and two-thirds jets. Fokker officials say they believe it may be possible to increase production to 50 aircraft in 1996, and Kroon said several large orders for the F70/100 combination are pending.

He added that leasing will play an increasingly important role in the market, and large leasing firms such as International Lease Finance Corp. could again buy Fokker aircraft as they have in the past.

Just how many of the aircraft produced will be F70s or F100s will depend on sales, and it will be possible to shift production targets with both aircraft moving down the same manufacturing line. The product mix can be changed with 14 months' advance notice. By 1996 this lead time will be cut to 12 months.

The two aircraft have 95% parts in common when both are equipped with the Rolls-Royce Tay 620 engine. The F100 can also be equipped with the more powerful Tay 650. Both aircraft have the some wing, with the exception of a few minor modifications on the F70. Kroon said that the company now has 27 firm orders and 11 options for the F70 since the program was launched at the Paris air show last year, plus three options for either the F70 or F100.

The first-series production F70 aircraft is in final assembly at Fokker's Schiphol plant here. As with the Fokker 100 work shore, large fuselage sections and the horizontal stabilizer are built by Deutsche Aerospace Airbus, while the wings are made by Bombardier Aerospaces's Short Brothers unit.

Two Rolls-Royce Tay 620 turbofan engines are to be mounted on the first-series production Fokker 70 in early July. When this aircraft is completed in mid-August it will be used to verify some of the test data gathered by the current prototype, the modified F100. Certification is expected to occur in October, with first delivery of a production F70 planned at the end of the year.

Orders and options to date include:

- Sempati Air of Indonesia, with 10 orders and five options, with deliveries starting in March, 1995. The airline currently operates a fleet of seven F100s and six Fokker 27 turboprops.
- Pelita Air Service, with five orders. The nonscheduled carrier has purchased F27, F28 and F100 aircraft and operated them on behalf of Indonesia's state oil company.

- Air Littoral of France, with five orders placed at the same time the airline ordered one more F100 to add to the nine it already operates.
- Mesa Airlines Inc. in the U.S., with an order for two F70s and an option for an additional six.
- British Midland, which become the first buyer to order F70s and F100s at the same time. Five F70s and four F100s will replace McDonnell Douglas DC9-10 and DC9-30 aircraft. The order included three options, and British Midland will be able to specify later whether these will be F70s or F100s.

Kroon said the British Midland order is significant in that the airline considered hush-kitting or reengining their aging DC9s, but then came to the conclusion that a combination of F70s and F100s was a better choice. The replacement market for aging DC-9s, Boeing 727s and Lockheed 1011s is a key one for the F100/70 family. Kroon said replacement of these aircraft at this time is being delayed due to the financial constraints.

The plan is to certify the F70 under a common type rating, and Fokker officials anticipate that difference training for flight crews may amount to as little as 30 min. in the classroom or in a simulator. The base of simulators in place now to support the fleet of 237 F100s delivered so far can be converted with software upgrades to serve both F70 and F100 pilot training needs. This will make it much easier to support the new 70-seat jet.

Kroon said the F70 should make it easier for turboprop operators wanting to graduate into jet operations to make the transition and provide a smaller-size option for larger carriers wishing to make the transition down from the larger F100 category. The F70 offers direct operating costs 18% lower on a per-trip basis than the F100, and Fokker officials contend it has the lowest structural weight per seat in its size class.

The F70 will come with Category 2 approach capability as a standard feature, with autoland as an option. The F70 will have the same core avionics systems as the F100, including EFIS and an automatic flight control and augmentation system. However, airlines will be able to specify a top-of-the line F100 avionics suite if they prefer. This would include a flight management system and Category 3 autoland capability.

Fokker is also selling an executive jet version of the F70, which can be configured to carry 30-52 passengers in first-class seating for distances of more than 2,100 naut. mi.

FOKKER 70 SPECIFICATIONS

POWERPLANTS

Powered by Rolls-Royce Tay Mk620 turbofan engines. The Tay 620 is rated at 13,850 lb. takeoff thrust and was certified by the British Civil Aviation Authority in 1986.

DIMENSIONS

Length	91.5 ft. (27.88 meters)
Height	27.9 ft. (8.51 meters)
Wingspan	92.1 ft. (28.08 meters)
Wing area	1,006.5 sq. ft. (93.5 sq. meters)
Wing aspect ratio	8.43
Cabin height	6.6 ft. (2.01 meters)
Cabin width	10.2 ft. (3.1 meters)
Standard cabin configuration	Five abreast, mixed 31- and 32-inch pitch, 79 passengers
Maximum seating capacity	79
Total luggage and cargo volume	762 cu. ft. (21.58 cu. meters)

WEIGHTS

Maximum takeoff weight	81,000 lb. (36,740 kg.)*
Maximum landing weight	75,000 lb. (34,020 kg.)
Maximum zero fuel weight	70,500 lb. (31,975 kg.)
Typical operating weight empty	49,984 lb. (22,673 kg.)
Maximum fuel capacity	17,065 lb. (7,740 kg.)

*The Fokker 70 has two optional weights. A 84,000 lb. MTOW aircraft has a range of 1,415 naut. mi. and a 88,000 lb. MTOW aircraft has a range of 1,840 naut. mi. The standard 81,000 lb. MTOW aircraft has a range of 1,080 naut. mi. An optional 984 U.S. gal. integrated center wing tank boosts total fuel capacity from 17,065 lbs. to 23,656 lbs.

PERFORMANCE

Takeoff field length at maximum takeoff weight (81,000 lb.), at sea level, no wind	4,250 ft. (1,225 meters)
Landing field length at maximum landing weight, sea level, no wind	3,970 ft. (1,210 meters)
Maximum cruise speed	Mach 0.77 (462 kts. true at 26,000 ft.)
Maximum range with 79 passengers*	1,080 naut. mi. (2,000 km.)

*For standard weight aircraft (81,000 lb. MTOW). See note above under Weights for ranges of optional MTOW aircraft.

Efficiency, automation combined on Fokker 100

Bruce D. Nordwall/Amsterdam
April 18, 1988

Fokker Aircraft has incorporated a high level of automation in the cockpit of its Fokker 100 transport to increase operational efficiency, greatly relieve flight deck workload during departure and arrival and provide a new level of protection against pilot error and system malfunctions.

This *Aviation Week & Space Technology* pilot was introduced to the Fokker 100 during automatic landing tests that were taking advantage of strong and gusty winds to gather data for crosswind certification. Fokker plans to seek certification for the landing system under Category 3B (automatic landing) conditions and high-speed rollout for a direct crosswind component up to 25 kt. The forecast crosswinds of more than 30 kt. made a good test likely.

Wim J. Huson, experimental test pilot for Fokker's flight test department, conducted the preflight briefing and rode in the right seat during the *Aviation Week* evaluation of the Fokker 100. The aircraft is designed for a two-man cockpit with a jump seat for check rides. Flight test engineer Wim Burgers occupied the jump seat to gather data and provide assistance.

We were flying the second prototype, which has a representative cockpit with only a few small differences from the production aircraft.

Fokker 100 flies over the North Sea during checkout prior to delivery to launch customer, Swissair. In the standard configuration the new Mach 77 aircraft is designed to carry 107 passengers for 1,305 naut. mi. Fokker is particularly proud of the economy of operation and low noise levels of the new aircraft.

The cabin area, like most prototypes, was not fitted with passenger seats, but contained instrumentation and systems to alter the center of gravity for tests.

Fokker obtained the basic type certificate on Nov. 20, 1987, from RLD, the Dutch airworthiness authority. Two prototypes and one production aircraft flew a total of 1,093 hr. to provide the supporting data.

For the first of two *Aviation Week* flights, Rudi Den Hertog, chief engineer for the Fokker 100, joined us with five technicians to operate the instrumentation package. The flight to gather data on crosswind landings was planned for about 1.5 hr. We would then return to the Fokker line, drop off the technicians, fuel to a representative fuel load, and depart for the evaluation flight.

After a quick walk-around inspection we boarded the aircraft. The prestart checklist in the cockpit has been simplified to eight items. Engine starting procedures are very straightforward. Huson started the No. 2 engine, and then I started the No. 1 engine. The turbine gas temperature peaked at 490C, well under the 700C limit. At idle and 52% N_2 (high-pressure turbine rpm), TGT was 350C and fuel flow indicated 330 kg. (728 lb.) per engine/per hour.

We taxied out with 7,450 kg. of fuel. Nosewheel steering is controlled with a conventional steering wheel on the pilot's outboard side console. The steering responded normally, but my initial application of the brakes would have spilled coffee in the cabin.

I had been told about the performance of the interim brake system during the briefing. Fokker first used carbon brakes. When a vibration problem developed, Loral, the brake manufacturer, produced an interim steel brake to be used while the carbon brake was redesigned. Fokker certified and delivered the first aircraft to Swissair with steel brakes.

Fokker improved the brake feel characteristics by adjusting brake springs. But the source of the problem is the brake control valve, which has been redesigned and will begin testing soon. Production aircraft will be delivered with carbon brakes and the new valve when testing is complete.

During design, Fokker worked closely with the airlines and the result is a very uncluttered cockpit. The dual Collins electronic flight instrumentation system (EFIS) displays at each pilot's console provide the primary flight instruments and a weather radar display. Airspeed, altitude and vertical speed are included on the primary flight instrument display, simplifying the instrument scan (AW&ST Feb. 23, 1987, p. 43).

Fokker uses the dark cockpit convention. Push-buttons that are functionally on are dark unless indicating a fault, and light up when a system is off. This contributed to the feeling of a clean cockpit.

One difference between the Fokker 100 I flew and U.S. aircraft is the on/off direction of toggle switches. Fokker determines the "on" direction of a toggle by assuming the pilot sweeps his hand forward from the back of the lower console, up the front console and back on the overhead. Any toggle shifted during that sweep would be turned on. Overhead toggles are the main difference, because toggles in the forward position are turned off in the Fokker scheme. But this is a matter of customer choice, Huson said.

The flight warning system also seemed well designed. It monitors engine and aircraft systems operation and provides alerts at four levels, depending on the urgency of the alert. Fokker uses the international standards adapted six years ago, Huson said.

A zero-level alert provides a memo, for information only, that is displayed as a blue-lit message on one of the two multifunction display units on the center main instrument panel. At the other extreme, a level-three alert—an emergency—is signaled by a repetitive triple chime, a red flashing master warning light and a red message on the display unit.

The flight warning system senses flight condition and inhibits alerts during phases of flight when they are not appropriate. All the immediate action and attention alerts are accompanied, on an adjacent multifunction display, by the procedures to follow. The implementation of the alert and warning system seemed particularly well thought out.

We completed the takeoff checklist during taxi, setting stabilizer trim at 1 deg. up and the rudder at 1 deg. right. Flaps would be set at 0 deg. for hot-and-high altitude takeoff, and at 15 deg. for an extreme

Prototype Fokker 100 gathered flight certification data in the good weather of Grenada, Spain. The lift dumpers on top of the wing and thrust reversers are shown deployed.

short-field takeoff, but we used the more normal 8-deg. setting. Maximum takeoff weight is 95,000 lb. (43,090 kg.), but our weight was 79,492 lb. (36,133 kg.). Our center of gravity would be 32.4% of the mean aerodynamic chord, close to the aft limit of 35%. Pressing the configuration check button simulated throttles forward and activated the takeoff warning system. Any discrepancy would have caused a level-three alert, but there was none.

The tower gave the wind as 270 deg. at 36 kt. as we lined up for takeoff on Runway 24 at Amsterdam's Schiphol Airport, checked controls and advanced the thrust levers. During takeoff the nosewheel steering is controlled by the rudder pedals, which give ±7-deg. nosewheel control.

Huson had previously computed and entered the takeoff parameters in the flight management system, which displayed V_1 and V_2 on the EFIS. Critical engine failure speed, V_1, was 122 kt., and takeoff/initial climb speed with one engine inoperative, V_2, was 128 kt. I rotated at V_r, which is equal to V_1 in the Fokker 100, and the aircraft lifted off immediately. Takeoff distance appeared to match the calculated performance of 4,200 ft. (1,380 meters) from the start of takeoff roll to 35 ft. altitude. We raised the gear and flaps and climbed out at 170 kt. Departure control cleared us to 2,000 ft. and provided radar vectors to set up for an automatic ILS approach and landing to Runway 19R.

I was impressed with the takeoff and climb performance of the aircraft. At maximum power setting for takeoff, the engines are each rated at 13,850-lb. thrust. The Fokker 100 was light and responsive on the controls and trimmed easily. After cleaning up the aircraft and reducing power from the takeoff engine pressure ratio of 169 to 154 for climb, our rate of climb was 3,000 fpm, and we were burning fuel at 4,840 lb. per hour per engine (2,200 kg.).

The changeable weather we experienced is typical for late winter, according to Huson. At various times blue sky, overcast, rain and hail blew rapidly through our area. Huson had informed me that this airplane was designed to be flown largely using the automatic flight control and augmentation system. AFCAS gives the aircraft a vertical as well as horizontal navigation capability. The pilot can choose either the most economical or most rapid climb, consistent with flight plan restrictions.

The reduction in pilot workload with the flight management system (FMS) coupled to AFCAS is significant and should increase flight safety, particularly in the high-density, high-workload part of the flight. The intermittent VFR conditions let me appreciate the extra time available to look for other traffic with AFCAS flying the aircraft.

Green-dot airspeed

We leveled off at 2,000 ft. and noted 110 lb. of fuel consumed since the start of the takeoff roll. We maintained a clean configuration to initial approach, flying 20 kt. above the airspeed displayed as a green dot on the airspeed display on the EFIS. The green dot airspeed is a reference airspeed, derived from the angle of attack, which the pilot can use to compute approach speeds independent of aircraft weight.

One of the few things a Fokker 100 pilot has to remember is to fly green dot plus 20 kt. for a clean aircraft on approach, green dot minus 10 kt. with 8-deg. flaps, and green dot minus 20 kt. for gear down and 25-deg. flaps. Green-dot airspeed for the first approach was 168 kt.

Inbound on the localizer, we dropped 8-deg. flaps and decelerated to 158 kt. when the ILS indicator showed the aircraft to be two

In the standard configuration, the Fokker 100 seats 107, five abreast, in a single class. Options range from a first class/economy mix of 97 passengers to a 119-seat high-density option.

dots below glideslope. At 1 dot before glideslope intercept, we dropped the landing gear, lowered the flaps to 25 deg., and slowed to 149 kt. as we started the descent. At 1,500 ft. we lowered the flaps completely to 42-deg. and slowed to 138 kt. Under calm wind at our weight the final approach speed would have been 128 kt.

The AFCAS held the airspeed smoothly during each transition. For each speed change we dialed in the change on the flight mode panel, and the autothrottle system made the correction. EFIS displayed green dot airspeed, stick shaker airspeed and 1.3 times stall speed, removing one more set of calculations from the pilot. Those workload savings should be particularly important to a pilot on short routes with many landings each day.

For the first approach to Runway 19R, the wind was still reported at 286-deg., 36 kt. But the winds varied in direction and strength, and I could see the flight controls and throttles working to keep the aircraft on glide path. Despite the strong crosswind gusts, the aircraft held the glidepath, crabbing above 150 ft. and switching to a wing-down and top-rudder crosswind correction at that altitude. At 50 ft. the system commanded a flare and reduced thrust levers to idle. My hands were very close to the yoke but no help was needed. Lift dumpers extended automatically and the aircraft slowed aerodynamically, rolling out on the centerline of the runway while still on the ILS course under autoland control.

Cleared for crosswind takeoff

Approaching the end at about 20 kt., I touched the brakes and stopped the aircraft. We were cleared for a 180-deg. turn on the runway and since the active runway was 27, were cleared for a takeoff on OIR, with more than 30 kt. of direct crosswind. With clearance for takeoff I rolled aileron into the wind, advanced the thrust levers to an EPR of about 1.30 and lifted the takeoff/go-around (TOGA) trigger on the throttles. TOGA automatically advanced the throttles to takeoff EPR of 1.69, removing one more bit of pilot workload, so I could concentrate on flying rather than setting the maximum thrust without exceeding EPR and TGT limits.

This time I engaged the autopilot as we lifted off and climbed out at the standard 18-deg. attitude. I felt more comfortable monitoring the aircraft performance and looking for traffic with the AFCAS flying the aircraft. We made three more ILS approaches to full-stop landing. I noticed a voice on the intercom announcing the aircraft altitude on final approach at 200 ft., 100 ft., 50 ft. and every 10 ft. to the deck, and asked Huson if it was a technician annotating a tape for the data-gathering flight. It was not. The verbal announcements are computer-generated by the ground proximity warning system.

As the aircraft touched down on each automatic landing, Huson called out the location of the touchdown point as compared with the target. A camera linked to an inertial system recorded the data precisely, but Huson's visual report was within 1 meter of the centerline on each landing, with winds gusting to as much as 38 kt. On-course touchdown appeared to be equally precise. During rollout the autoland held the aircraft to within 1 meter on the downwind side. Five approaches satisfied the data requirements, and we returned to the Fokker line to drop off the technicians and refuel.

Ship two, the prototype we flew for both flights, has a greater zero fuel weight than production aircraft because of the instrumentation installed. We added fuel to 16,953 lb. (7,690 kg.), which Huson said was representative of the weight of an aircraft with a full load of passengers, fueled for a short flight such as Paris to Zurich, with fuel reserve for a half-hour divert and 45-min. holding. Fokker used an advanced transonic airfoil concept with some aft loading, which provides a higher critical Mach number for relatively more wing thickness, according to Fokker. The high-aspect-ratio wing is longer than the F-28 4000's, producing less induced drag and more efficiency. The lower drag at cruise altitude gives the aircraft better maximum cruise speed. The certification change to permit maximum cruising speed to increase from Mach 0.75 to 0.77 will be accomplished by the end of the year, Huson said.

Low operating cost

The use of composites, adhesive metal bonding and efficient construction give the Fokker 100 the lowest operating empty weight per seat of any jet up to 160 seats, according to the company. The result is that the Fokker 100 will have the lowest operating cost per aircraft mile of any current jet transport. The break-even passenger load for the 107-seat configuration is 42 passengers, the company said.

With our new 79,200-lb. (36,000-kg.) takeoff weight, the display indicated the same V_1 and V_2 as for the first flight. For this flight we would use Runway 27, the duty runway, and with a 25-kt. headwind, computed a takeoff field length of 3,214 ft. (980 meters).

Takeoff performance matched our computations, and we followed departure control instructions for a vectored departure to the northwest. Huson had preset the way points en route to the operating area over the North Sea. On the way out I activated the Bendix RDR-4A color weather radar. I found the radar operation straightforward and the presentation easy to interpret. The coastline showed up well, too, in the map mode. Fokker designed the radar interfaces so that the customer can select any ARINC 700 system.

We leveled off at 20,000 ft. 9 min. after takeoff, having burned 1,278 lb. (580 kg.) of fuel. I flew the Fokker 100 through a series of turns, increasing the bank to 60 deg. The aircraft trimmed well, was stable in pitch and responsive laterally.

For the first turns I used a conventional scan of the attitude, vertical speed and altitude presentations on the primary flight display. Huson then suggested that I try using the flight path vector, a symbol on the attitude direction indicator that displays the aircraft's flight path. I found it to be another useful innovation.

After a clearing turn I throttled back and eased the nose up to enter a wings-level, power-off stall. The autothrottle was not engaged at this point, but at about 145 kt.—about 1.3 V_{stall}—the throttles advanced automatically. Huson explained that this demonstrated one of the many protections designed into the system. The AFCAS logic directs this action based on measured angle of attack, and makes an unintended stall very unlikely, Huson said.

To permit me to evaluate the stall characteristics, Huson had to switch off two push buttons in the overhead console to disconnect the autothrottle system. This caused a flashing "autothrottle" alert on the flight mode annunciator, and the word "manual" to appear in amber in the speed window on my primary flight display, a clear reminder that in the Fokker 100, manual operation is an unusual situation.

I reduced the throttles to idle and established a 15-deg. nose-high attitude. The stick shaker came on at 119 kt., with the onset of stall buffet at 112 kt. I continued holding back pressure on the yoke, with level wings, and the aircraft stalled at about 110 kt., with a gentle roll off on the left wing and nose drop. By leveling the wings and adding power, the aircraft recovered immediately.

I shifted to a dirty configuration with gear down, 42-deg. flaps and idle power, and eased the nose up. The stall shaker came on at 104 kt. with stall buffet at 94 kt. This time there was more vibration while decelerating from stall shaker to stall, and at the stall the aircraft rolled off and dropped its nose more sharply. Recovery was positive once again, with good control response, and the aircraft flew out of the stall.

Simulating a 3-deg. glideslope approach to a high-altitude airfield, such as La Paz, Bolivia, I entered a 700-fpm rate of descent with gear down, under AFCAS control. At Huson's suggestion, to make it even more demanding, we used full flaps—42 deg.—instead of the usual 25 deg. At 12,000 ft. I started a go-around, pulling the TOGA trigger. After establishing a positive rate of climb of 2,000 fpm we raised the flaps to 15 deg. and raised the landing gear.

Pulling back one thrust lever, the aircraft climbed at 1,000 fpm, but I probably was slowly bleeding off airspeed and did not wait long enough to achieve a steady-state climb. A clean aircraft on one engine and at that weight and altitude would normally climb about 650 fpm.

The protection against pilot error intrigued me, and I slowed the aircraft in the landing configuration to look at another situation. In most aircraft if the flaps are inadvertently raised at too slow an airspeed and the pilot maintains altitude without adding power, the aircraft will decelerate and stall.

Automatic throttle advance

I tried it in the Fokker 100, and as soon as the flaps started coming up, the protection system again automatically advanced the throttles. The system is designed to advance the throttles to give the aircraft 15-kt. margin above stick shaker speed, Huson said. The aircraft may not be completely "pilot-proof," but it has more protection than previous aircraft in this size category.

As we returned for some touch-and-go landings on Runway 27, the sun had set and the weather was reported as 2,500 ft. scattered, variable broken, visibility 5 mi. and winds from 290 deg. at 24 kt. We were given radar vectors to a straight-in visual approach to Runway 27. The engines were so quiet that I could not tell from their sound how much of a power change I made when moving the throttles. Without that feel, I found it necessary to refer to the engine instruments to make sure the power adjustments on the glideslope were as intended. With a bit more time in the seat, that would be overcome. The speed trend vector on the primary flight display, which continually predicts what the speed will be over the next 10 sec., is another helpful addition.

Speed brakes

The aircraft responded well and felt stable during the touch-and-go landings, but my leisurely flare caused me to land a bit long on the first one. After Huson reset the trim and raised the flaps to 15 deg., I advanced the throttles and rotated to 18 deg. We were the only aircraft in the pattern, and rate of climb permits a very tight pattern. We arrived at 1,500 ft. agl. while in the turn and transitioned to landing configuration for the turn to base with very little downwind leg.

In fact, I had to work to avoid going too fast. I used the tail-mounted speed brakes several times, since they can be used in flight and have no noticeable effect on pitch or stall. The speed brakes are the same as those on the F-28, but do not have variable control. The Fokker 100 system selects only full-in or full-out, with an airspeed bleed back at greater than 200 kt.

The second touch-and-go felt good as I flew a better pattern. Just after takeoff Huson pulled the left throttle to idle as we lifted off. No large rudder correction or trim was required while I cleaned up the aircraft and turned right downwind into the good engine. I flew a slightly wider and longer pattern but the aircraft handled well, with sufficient power and control to prevent the gusty winds from becoming a problem.

I shot two more landings—one with no flaps at an approach speed of 150 kt., and another by hand-flying a crosswind landing. My crosswind approach and landing was not as smooth as the autoland's on the previous flight, but the aircraft was comfortable in both the crab and the slip approach.

Huson took the controls to demonstrate short-field performance for the full stop landing. We armed the lift dumpers, so they would automatically deploy on touchdown. Flying at normal approach speed, he touched down and used maximum braking. The aircraft has an antiskid system, but I had not challenged it with full pressure. With maximum braking Huson stopped the 70,400-lb. (32,000 kg.) aircraft in 984 ft. (300 meters) from touchdown without using reverse thrust.

Taxiing back in, we stopped on a wide part of the ramp to back up with reverse thrust. Lifting and pulling the reverse levers mounted on the throttles, I added a little power and the aircraft taxied backwards. With the aft center of gravity, I stayed off the brakes to avoid any chance of tip back, and came out of reverse thrust to stop. Huson pointed out that operators would need individual approval from the FAA, Fokker and Rolls-Royce before using reverse thrust for taxi.

Aircraft noise levels

There were several features of the Fokker 100 that I did not get to observe. One was the noise level. The cockpit in the prototype I flew had a comfortable noise level, but production aircraft are much quieter, according to Huson. Fokker test data show that exterior noise for both the Tay and 650 engines average an effective perceived noise level 5 dB less than the FAR 36 Stage 3 requirements for community noise.

The standard configuration is for 107 passengers in a single-class, one-aisle seating with five abreast at 32-in. seat pitch. Other options include a 97-passenger first-class/economy configuration—12 first class seats at 36-in. pitch and 85 economy seats at 32 in. A 105-passenger business/economy combination consists of 55 business-class seats at 34-in. pitch and 50 economy seats. A high-density option has 119 seats.

All new aircraft look good, but the Fokker 100 I saw showed great attention to detail. The aircraft seemed to reflect pride in workmanship by both the production and maintenance crews at Fokker.

Fokker has orders for 87 aircraft and options for 91 more. Swissair, KLM and USAir are the launch customers, with the first Swissair aircraft delivered Feb. 29. KLM is scheduled to receive its first aircraft in September and USAir in the first quarter of 1989.

The first North American operator will be Inter-Canadian, which will lease four aircraft from International Lease and Finance Corp., Los Angeles. Although not considered a launch customer, International Lease and Finance will receive the second seven production aircraft. InterCanadian is scheduled to receive its first aircraft this fall. Future growth plans include increasing maximum takeoff weight in two steps, first to 98,000 lb. (43,090 kg.) and then to 104,000 lb. (47,272 kg.).

FOKKER 100 SPECIFICATIONS

POWERPLANTS

Powered by Rolls-Royce Tay 620 or 650 turbofan engines. The Tay 620 is rated at 13,850-lb. takeoff thrust, and the 650 is rated at 15,100 lb. The Tay 620 was certified by Britain's Civil Aviation Authority in 1986, and the 650 is scheduled for certification in August.

DIMENSIONS	
Length	116.6 ft. (35.53 meters)
Height	27.6 ft. (8.41 meters)
Wingspan	92.1 ft. (28.08 meters)
Wing area	1,006.5 sq. ft. (93.5 sq. meters)
Wing aspect ratio	8.43
Cabin height	6.6 ft. (2.01 meters)
Cabin width	10.2 ft. (3.1 meters)
Standard cabin configuration	Five abreast, 32-in. pitch, 107 passengers
Maximum seating capacity	119 passengers
Cargo compartments	606 cu. ft. (17.16 cu. meters)

WEIGHTS Standard (Tay 620)	
Maximum takeoff weight	95,000 lb. (43,090 kg.)
Maximum landing weight	88,000 lb. (39,915 kg.)
Maximum zero fuel weight	81,000 lb. (36,740 kg.)
Typical operating weight empty	53,695 lb. (24,355 kg.)

Optional (Tay 650)	
Maximum takeoff weight	98,000 lb. (44,450 kg.)
Maximum landing weight	88,000 lb. (39,915 kg.)
Maximum zero fuel weight	81,000 lb. (36,740 kg.)
Typical operating weight empty	53,975 lb. (24,484 kg.)
Maximum fuel capacity	17,500 lb. (3,445 gal.)

PERFORMANCE

Takeoff field length at maximum takeoff weight, sea level, no wind, Tay 620 engine	6,560 ft. (2,000 meters)
Landing field length at maximum landing weight, sea level, no wind, Tay 620 engine	4,560 ft. (1,390 meters)
Maximum cruise speed	Mach 0.77 (450 kt. true)
Maximum range with 107 passengers, Tay 620 engine	1,305 naut. mi. (2,417 km.)

Fly-by-wire, digital avionics ease A320 transition training

David M. North/Toulouse
November 30, 1987

Airbus Industrie has achieved its goal of integrating the A320's advanced technology, including the fly-by-wire flight control system, side-stick controller and digital avionics. The aircraft's cockpit design and stable handling characteristics provide an easy transition for pilots accustomed to more conventional aircraft.

An evaluation flight by this *Aviation Week & Space Technology* pilot in the No. 1 A320 prototype at the Airbus facility here convinced me that Airbus has carefully matched human engineering and pilot workload with aircraft performance to develop a transport that is enjoyable to fly. The company, in this process, has not forgotten possible concerns over an almost fully electronic cockpit and automatic safety features, and has provided a redundancy in systems that should be able to overcome almost any combination of emergencies.

Standard cabin configuration for the A320 is for 12 seats with a 36-in. pitch in first-class and 138 seats in the economy section with a 32-in. pitch. A high-density configuration includes 176 seats with a 30/29-in. pitch.

The technology advances of the A320 are more readily seen in the cockpit. But Airbus also has used composite materials in the construction of the 150-seat aircraft, equipped the aircraft with turbofan engines commanded by digital fuel controls and provided commercial operators with a large cabin with flexibility in seating arrangements.

The flight in the No. 1 prototype was with Pierre Baud, vice president of the Airbus flight division. The aircraft was equipped with icing test equipment, so we were limited to a maximum speed of 300 kt. The speed limitation was imposed to prevent damage to test equipment, mounted at different locations on the outside of the aircraft. The equipment also produced some drag, slightly increasing the fuel flows registered during the flight.

Left seat rotation

Four pilots, including myself, took turns flying the A320 from the Toulouse airport. Baud sat in the right seat during the flight while the four pilots rotated through the left seat. All four flew the aircraft at cruise altitude to familiarize themselves with its slow-flight characteristics, and a number of approaches, landings and takeoffs were made by each to gauge performance in the critical flight regimes.

Following a briefing by Baud on the flight characteristics of the A320's fly-by-wire control system, the aircraft was pre-flighted. The weight of the aircraft on the ramp was close to 120,000 lb., or 76% of its maximum 158,700-lb. takeoff gross weight. There were four Airbus personnel in the aircraft and flight test instrumentation.

The fuel on the aircraft was 22,800 lb., or little more than half the maximum available. The standard configuration with fuel in the wings has a maximum capacity of 27,202 lb., but an optional tank in the wing center section raises maximum capacity to 41,470 lb.

Baud calculated the V_1 takeoff decision speed to be 130 kt. and the rotation speed to be 139 kt. The takeoff safety speed was 141 kt., and also was portrayed on the multipurpose control and display unit on the center pedestal.

Uncluttered work area

The first impression of the A320 cockpit is of an uncluttered work area with few controls and monitoring instruments. The overhead panel is broken into two sections with aircraft system control switches on the front of the panel and circuit breakers to the rear. The center pedestal contains two multipurpose control and display units, the radio management panel, radar control, throttles and speedbrake and flaps handles.

The cockpit area is wide and one can easily slip into the pilot seats without the contortions required in some transport aircraft. This ease of entry is aided by the absence of a control column. The left armrest in the pilot's seat is adjustable so that the left arm can rest easily and facilitate wrist movement in grasping the side-stick controller.

Both the General Electric/Snecma (CFM International) CFM56-5A1 turbofan engines were started in the automatic mode by simply depressing the main start buttons on the overhead panel. The engine parameters were monitored on the top display unit on the center instrument panel.

The limiting thrust parameter for the CFM International turbofan engines is N_1 low-pressure compressor speed. The limiting parameter for the International Aero Engines V2500 will be engine pressure ratio (EPR). The 25,000-lb.-thrust V2500s are planned for aircraft to be delivered in mid-1989. The limiting thrust parameters are computed by the full-authority digital engine control for all phases of ground and flight operations.

The exhaust gas temperature on both engines peaked well below the maximum limit on the 16C day and was displayed on the control

Airbus Industrie A320 is equipped with sidestick controllers for pitch and roll commands through the fly-by-wire control system. The instrument panel contains six 7.3-×-7.3 in.-display units. Primary flight and navigation displays are located in the front of the pilots.

Air France is one of the first airlines scheduled to receive the A320. The airline has ordered 25 of the transports.

unit. At idle, the exhaust gas temperature (EGT) was 445C and the fuel flow was close to 700 lb./hr. The slats were positioned at 22 deg. and the flaps at 20 deg., the normal takeoff configuration.

Both pilot stations have a nose-wheel steering wheel to the outside of the seat. The nose-wheel steering was sensitive to turn commands, with some overcontrol possible at first. Baud said that more friction would be used in the system in later aircraft. The brakes also grabbed at slow speeds, a condition that Baud said would be corrected in production aircraft.

Takeoff in the A320 was accomplished by one of the other pilots 11 min. after the beginning of taxi; I occupied the jump seat. A total of 640 lb. of fuel had been consumed since engine start. The takeoff roll was accomplished in less than 4,000 ft. when the nose of the aircraft was raised to achieve a 18-deg. pitch attitude. The takeoff was smooth and acceleration brisk throughout the takeoff and raising of the landing gear and flaps.

The initial fuel flow after takeoff was 8,290 lb./hr. at 180 kt. The fuel flow dropped to 7,060 lb./hr. passing through 5,000 ft. while indicating 220 kt. It required less than 5 min. to reach 10,000 ft., where the fuel flow was 6,610 lb./hr. Less than 1,500 lb. had been used since takeoff.

Maximum altitude reached

At 20,000 ft., the rate of climb was close to 3,000 fpm and the fuel flow was 5,290 lb./hr. A total of 10 min. had been required to achieve 20,000 ft. and another 2 min. was taken to reach 25,000 ft. A total of 2,500 lb. of fuel was required to reach 25,000 ft., the highest altitude to be attained during the flight. The A320 was stabilized at this level at 280 kt. and the fuel flow was 3,100 lb./hr.

Airbus Industrie estimates that at a maximum design takeoff weight of 162,040 lb. for later A320s, it would require 3,910 lb. of fuel, 26 min. and 167 naut. mi. to reach a cruising altitude of 35,000 ft. At a takeoff for a typical mission of 500 naut. mi., the fuel required would be 3,480 lb. and a total time of 23.5 min. would be required to reach 37,000 ft.

Estimated cruise data for a production A320 shows a maximum speed of Mach 0.82 and a fuel flow of 6,890 lb./hr. at 28,000 ft. The long-range cruise speed for the A320 at 37,000 ft. is Mach 0.79, or a true airspeed of 455 kt. The fuel flow at the economic cruise speed is 4,760 lb./hr.

Airbus design philosophy dictates that all aircraft system lights are out in the overhead panel under normal conditions, a practice followed with the A320. During normal cruise operations, the primary engine indications are shown on the upper control and display unit.

The same display contains more than 25 memo items applicable to the operation of the aircraft, such as the no-smoking light on, speed brakes extended and fuel-feed sources. The lower display normally contains a system page showing schematics of the fuel, electric, flight control, air-conditioning, hydraulic and other systems. A fault in a system will show on the upper display unit along with the corrective action, while the schematic indicating the fault is displayed on the lower screen.

Once in the left seat, I found that the A320 pilot seat was comfortable and the visibility excellent, including the ability to see the left wingtip without much effort. The side-stick controller was easy to grasp with the left arm resting on the extended arm rest. Five minutes at the most was required to become accustomed to the side-stick con-

Airbus Industrie has accumulated more than 900 hr. of flight testing on four A320s since the aircraft first flew on Feb. 22, 1987. The A320 is powered by two CFM International CFM56-5A1 turbofan engines. Air Inter, which also will be one of the first carriers to receive the A320, has placed orders for 12 of the transports. In addition, Indian and Royal Jordanian airlines will receive A320s.

troller, which differs from the control column installed in most transports and the stick found in fighters.

My most recent experience was flying the USAF/Rockwell International B-1B bomber and corporate-type jet aircraft. Two of the other pilots in the group, a retired British Airways Lockheed L-1011 captain and a current captain for Air Inter flying the Caravelle, easily adapted to the A320's side-stick controller.

A number of turns were made in both directions at 25,000 ft. to become accustomed to the controller and the instrument scan required to monitor the response of the aircraft. It was easy to read the primary flight display and assimilate the flight information portrayed. The aircraft's airspeed and altitude were shown in a vertical format. The rate of climb indicator has a needle that is hinged on the right and points to numbers on the left, and this took a little time to read correctly.

During this portion of the flight, Baud explained the control laws used in the fly-by-wire flight system in the A320 (AW&ST Sept. 22, 1986, p. 40). On the ground, there is a direct relationship between stick input and elevator and roll control surfaces. The aircraft transitions to C* control laws gradually within 5 sec. after takeoff. The C* control law for aircraft stability developed by the U.S. National Aeronautics and Space Administration is used throughout the flight regime and until 50 ft. from landing.

Control law provisions

The C* control law dictates that stick input commands a change in vertical load factor in pitch. The law provides:

- Short-term platform stability with the stick in neutral.
- Auto-trim function and neutral static stability.
- Aircraft response almost unaffected by speed, weight or center of gravity location.
- Bank angle compensation up to 33 deg.

At first, I kept looking for the pitch trim switch normally located at the top of the control column to compensate for pitch changes. It did not take long to realize that the flight control computer was continually compensating for pitch changes and further input was not required. This also was true for any rudder inputs that might be required to compensate for induced yaw.

The C* law provides for high-speed protection in flight. Baud explained this protection in light of the A320 configuration. If maximum Mach or speed is exceeded, the elevator receives an automatic pull-up command, the bank angle is reduced to 40 deg. and pilot nose-down authority is reduced. The maximum stabilized speed with full nose-down stick is the maximum speed plus 16 kt. The maximum achievable speed with full control is maximum speed plus 32 kt. Maximum bank angle is 67 deg. inside the normal flight envelope, and when removing control input the aircraft will automatically revert to a 33-deg. bank. Baud said that while exceeding the maximum speed, the flight control computer will first slowly level the wings before commanding a nose-up action to reduce the speed below the maximum limits.

While still at altitude, Baud directed the flight test engineers in the cabin to simulate failures in the flight control system. With a double failure in a number of the flight control computer systems, the aircraft reverts to both pitch and lateral alternate laws with protection.

The main difference in this highly unlikely situation of multiple failures is that pitch and roll protection is not available. The high-speed protection is replaced by positive static stability, which may be overridden by the pilot. High-angle-of-attack protection is replaced by low-speed stability and an aural warning of stall is still available.

Further failures in a combination of the flight control computers reverts the aircraft to pitch- and lateral-direct laws, where elevator, aileron and spoiler deflections are proportional to the stick commands. Flying in this mode, the control of the A320 was positive and not unlike normal flying characteristics of other commercial transports.

Baud then introduced a full electrical failure to the flight control system, including the ram air turbine that is used to supply electrical power in the case of multiple generator failures. The instrument panel was still functioning at this point. I was easily able to command up to 30-deg. banked turns while descending by the use of rudders for roll and the large pitch trim wheel on the center pedestal for pitch control. Baud said that a landing with a complete electrical failure in the A320 was not difficult, and that he had accomplished some during the flight test program.

The "C*" control law provides load factor protection of 2.5g in the clean configuration and 2g when the flaps are down. During slow flight conditions the control law allows for angle-of-attack protection. At 11,000 ft., Baud requested that I raise the nose to 30 deg. at slightly over 200 kt. The throttles were set for automatic response throughout the entire flight.

The stick was full aft at 140 kt. and the attitude had dropped to 20 deg. As the speed decreased to 105 kt. the engine power advanced to 94.8% N_1 and rate of climb of 800 fpm was established with a near 25-deg. nose-up attitude. The throttles themselves do not advance during the maneuver once placed in one of the automatic settings in the throttle quadrant.

The automatic stall protection in the A320 activates at 3-5 deg. below the stall angle. During the stall sequence the aircraft was rolled to the maximum of 45-deg. available. The flight characteristics of the A320 during the slow-flight maneuver were excellent, and there was no uncomfortable feeling operating at low speeds and high angles of attack. The pitch attitude at the low speeds is limited to 25-deg. nose up and 15-deg. nose down. At the higher speeds, the nose-up pitch is limited to 30 deg.

This same maneuver would be used to combat wind shear encountered during the landing phase in the A320. As the pilot pulls the side-stick controller full back when encountering wind shear, the autothrottle commands go-around power and the flight control computers automatically configure the aircraft for maximum lift. If further studies into combating wind shear reveal this not to be the best corrective action, Airbus can modify the flight control computers for the best results. A simulated wind shear encounter on one approach into Toulouse airport resulted in full power and a 1,600-fpm climb at close to a 25-deg.-pitch attitude.

During the two instrument landing approaches I made to Runway 15 at Toulouse, the A320 was responsive in both pitch and roll and there was no tendency to overcontrol with the use of the side-stick controller. A pilot flying the A320 in normal commercial operations will have difficulty returning to another transport aircraft with less advanced technology because of this aircraft's ease of operation and splendid instrument panel.

The pattern at Toulouse was flown at 160 kt. and at 2,000 ft. The fuel flow at this speed was close to 3,100 lb./hr. Deploying the flaps and slats did not cause any change of pitch, and there was little notice of their retraction or deployment except in the diagram in the top display unit. The landing gear was lowered at 140 kt., again with no noticeable pitch change. The glide slope was flown at close to 125 kt. and there was little change in engine power to stabilize at the approach speed.

Below 50 ft., a landing law has been introduced into the flight control computers. The landing law is used to allow for a conventional flare. At 50 ft. a stick input commands a pitch attitude increment to a reference attitude. At 30 ft., if there is no change in command from the side-stick controller, the aircraft will assume a 2-deg.-lower pitch attitude. This allows for the pilot to pull back on the stick with the proper feel properties to command a flare. The first landing was firm but not hard. A go-around for the next approach was made, with quick acceleration on the runway and a 18-deg. nose-up attitude.

Following the next landing—a smoother one—Baud retarded the right engine to idle immediately after liftoff. The rudder trim of approximately 4 deg. was automatically introduced, and I was able to hold 15-deg. attitude with the left engine at full power and climbing at 600 fpm. At this point, Baud suggested I remove my hand from the side-stick controller, and the A320 retained the same attitude until reaching 2,000 ft. when a turn was commanded for the next downwind leg.

Control technique

Throughout the landing approaches, I found that holding the side-stick controller at the top with two fingers and a thumb was less likely to lead to attempts to overcontrol the aircraft. Grasping the controller with the entire hand was not necessary for positive control.

A total of 18,160 lb. of fuel was used during the 3.3-hr. flight. The majority of the flight was flown in the traffic pattern with a total of nine approaches and landings.

Although not demonstrated during the flight, the A320's right control computers have a load alleviation function. At speeds greater than 200 kt. and with slats retracted, the No. 4 and No. 5 spoilers and ailerons are symmetrically deflected up with a high rate upon detection of load gusts. Elevator input also is used to compensate for pitching moment induced by the spoilers and ailerons.

Airbus has accumulated more than 900 hr. of flight time on the four A320s used in the flight test program. Almost 830 hr. of the total flight time has been directed to Federal Aviation Administration certi-

Airbus Industrie has received 287 orders for its A320, with Northwest Airlines planning to acquire 100 aircraft. Airbus will attain a production rate of four aircraft a month in 1988 and expects to build to eight aircraft monthly in 1990.

fication requirements. Baud estimates that more than 70% of the A320 development has been accomplished and 40% of the FAA requirements have been met. Almost 75% of the required performance figures for the A320 have been completed.

Baud said the airline pilots who have flown the aircraft have generally high praise for the A320 and its handling characteristics. The certification of the A320 is planned for late February or early March, 1988.

The A320 has a standard configuration of 150 seats, including 12 seats in first class and 138 seats in an economy section with a 32-in. pitch. A high-density configuration for the A320 includes 176 seats with a 30-in. pitch. Airbus recently completed a test emergency evacuation of 179 people from the A320 in 81 sec.

Air France and British Caledonian are scheduled to receive the first A320s in April, 1988. The concurrent deliveries are to be followed by the delivery of the No. 6 A320 to Air Inter. As an indication of the A320's appeal to commercial operators, Airbus has recorded 287 firm orders for the transport before its certification. The addition of options for the aircraft raises the total to above 460.

"We have been talking with all of the operators holding options on the A320 to see whether we can locate some earlier delivery positions for firm orders, but the operators are all holding on to their options," John Leahy, vice president of marketing and sales for Airbus Industrie of North America, said. "We have three delivery positions available in the fourth quarter of 1991, and it is a drawback in attempting to sell aircraft to offer such a long lead time. Many operators in the U.S. are used to much shorter waiting times, but it is a tribute to the A320 that we have such a backlog."

Northwest Airlines has orders for 100 A320s and is scheduled to receive the first aircraft in June, 1989. Pan American World Airways is scheduled to receive the first of its 16 A320s at the same time as Northwest. Pan American has another 34 A320s on option. The Pan American A320s are to be powered by IAE V2500 engines.

Aircraft financing arrangements

"We are accused of working financial deals for our aircraft, but that is not correct," Leahy said. "Our arrangement with Northwest does not include any finances provided by us, and with Pan American we are only providing financing guarantees for their 16 aircraft on order. There is not a similar arrangement for their 34 aircraft on option."

"With the service flag being raised by the airlines, the A320 offers a lot of flexibility," he said. "The A320 is 7 in. wider than a Boeing 757, and with a wide-aisle option, the ability to move people around a service cart is a big factor."

The full retail price for the A320, equipped with seats, galley, lavatories, overhead bins and standard amenities is close to $35 million, according to Leahy. "We may be higher in price than the competition," he said, "but I believe we offer a high-technology aircraft that will cover almost any segment flown, except for the long-range route."

AIRBUS INDUSTRIE A320 SPECIFICATIONS

POWERPLANTS

Powered by CFM International CFM56-5A1 or International Aero Engines V2500 turbofan engines with a 25,000 lb. thrust rating each at takeoff. The CFM56-5A1 engines are certified and the V2500 engines are planned for aircraft to be delivered in mid-1989.

DIMENSIONS

Length	123.3 ft. (37.6 meters)
Height	38.6 ft. (11.8 meters)
Wingspan	111.3 ft. (33.9 meters)
Wing area	1,318 sq. ft. (122.4 sq. meters)
Wing aspect ratio	9.39
Cabin height	7.3 ft. (2.22 meters)
Cabin width	12.1 ft. (3.69 meters)
Standard cabin configuration, two classes	150 passengers
Maximum seating capacity	179 passengers
Cargo capacity below cabin floor	7 LD3-46 containers (1,407 cu. ft.)

WEIGHTS

Maximum takeoff weight	158,700 lb. (72,000 kg.)
Maximum landing weight	138,900 lb. (63,000 kg.)
Maximum zero fuel weight	130,100 lb. (59,000 kg.)
Typical operating weight empty	85,800 lb. (38,900 kg.)
Maximum fuel capacity	41,470 lb. (6,190 gal.)

PERFORMANCE

Takeoff field length requirement for a 1,000 naut. mi. segment with 150 passengers and baggage, sea level	4,700 ft. (1,433 meters)
Landing field length requirement for same trip	4,400 ft. (1,341 meters)
Maximum cruise speed	Mach 0.82 (482 kt. true)
Long range cruise speed	Mach 0.79 (455 kt. true)
Maximum range with 150 passengers and baggage	3,000 naut. mi.

737-300 shows added performance, workload cut

Robert R. Ropelewski/Seattle
November 12, 1984

U.S. domestic airline crews scheduled to begin operating Boeing's new 737-300 twin-engine transport in commercial service next month will find a strong commonality with the existing Boeing 737-200 combined with improved performance from higher-thrust engines and reduced workload from an advanced automatic flight control system.

Officials of the 737-300 program at Boeing expect to receive Federal Aviation Administration certification for the aircraft on Nov. 14, and initial deliveries are scheduled for the end of this month to USAir and Southwest Airlines. The two airlines will put the aircraft into scheduled commercial service shortly thereafter.

This *Aviation Week & Space Technology* pilot recently flew one of the 737-300 test aircraft from Seattle's Boeing Field. The 1.7-hr. flight included a trip to Moses Lake/Grant County Airport in eastern Washington state, where several automatic approaches, landings and go-arounds and two manual approaches and landings were performed. The flight provided ample opportunity to observe the CFM International CFM56-3 engines that power the 737-300 and the comprehensive flight management and flight control systems incorporated in the aircraft.

The 737-300 is a stretched version of the 737-200 incorporating a 44-in. fuselage plug just ahead of the wing, a 60-in. plug just aft of the

Boeing 737-300 flight test aircraft in USAir colors lands at Boeing Field in Seattle at the end of the Aviation Week & Space Technology *demonstration flight. USAir will be the first airline customer to take delivery of the 737-300, scheduled for late this month.*

wing, wing and horizontal stabilizer tip extensions and a modified dorsal fin. It is powered by two 20,000-lb.-thrust CFM56-3 high-bypass turbofan engines replacing the 16,000-lb.-thrust Pratt & Whitney JT8D-17 turbofan engines used on the most recent versions of the 737-200. Improvements in available thrust and in overall flight management of the 737-300 are quite distinct.

Boeing took a preliminary step toward the 737-300 flight controls configuration in the early 1980s when it developed and certificated an automatic flight control system with a Category 3A automatic landing capability for the 737-200 and began offering it as an option on that aircraft (AW&ST July 5, 1982, p. 57). That system, which has since been adopted by a majority of 737-200 operators, offers a performance computation capability but not the vertical and lateral guidance functions provided by the newer system.

The 737-300 flight management system carries the 737-200 automatic flight control system a step further by integrating all of the systems necessary to provide the functions of automatic flight control, performance management, precision lateral and vertical navigation and systems monitoring.

The flight management computer system used in the 737-300 is similar operationally to the system used in both the Boeing 757 and 767. It contains a periodically updated navigation data base that allows the flight crew to select preprogramed airline routes or direct routes and to enter any additional waypoints they desire.

Primary elements of the new aircraft's flight management system are:

- Lear Siegler flight management computer and control/display unit (CDU).
- Sperry Flight Systems SP300 digital flight control system (autopilot) and integrated mode control panel.
- Smiths Industries full flight regime autothrottles.
- Honeywell dual ring laser gyro inertial reference system.
- Digital-analog adaptor to perform the signal conditioning needed to interface the flight management system with the 737's standard Arinc 500 series avionics.

Although the flight management system uses only a single autopilot, the SP300 autopilot comprises two independent computing and monitoring channels intended to provide a fail-passive capability during automatic approaches and landings. If a fault is detected during an automatic landing, both channels are programed to disengage, leaving the aircraft control surfaces in trim. For failures occurring in other phases of flight, Boeing has had to demonstrate to the FAA that the aircraft can endure full hardovers in any axis long enough for the crew to disengage the autopilot and take over manually.

There are few differences between the cockpits of the new aircraft and its predecessor, except for the inertial reference system control/display unit in the overhead panel of the 737-300 cockpit and the somewhat larger size of the flight management system keyboard and display on the center console compared with the performance data system control/display unit of the 737-200.

This was intentional, because Boeing's objective is to insure that crewmembers already type-rated in the 737-200 automatically will be qualified to fly the 737-300 as well. To achieve this, the 737-300 will have operating characteristics similar to the 737-200 in almost all areas, including takeoff, rejected climb, obstacle clearance, approach, landing and go-around procedures, takeoff and landing flap settings, flap retraction and extension schedules, maximum altitude for flap extension, and landing gear and flap placards.

The two aircraft will have similar performance characteristics in such areas as stall speeds (all flap settings), operating speed (takeoff, maneuver, climb, cruise, approach, landing and go-around), field length limits for both takeoff and landing, climb capabilities with both one and two engines operating, takeoff flight path, terrain clearance, maximum altitude and descent capabilities.

Third aircraft

I occupied the left seat in the two-crew cockpit of the 737-300 for the *Aviation Week* evaluation flight in the aircraft, while Boeing project test pilot James C. McRoberts took the right seat. The aircraft used was the third of three to fly in the 737-300 flight test program, and its cabin was still filled with instrumentation and equipment rather than standard commercial furnishings.

Zero fuel weight of the aircraft prior to start was 76,338 lb., and we were carrying 24,100 lb. of fuel, for a total weight at engine start of approximately 100,400 lb. The 737-300 will be certificated initially at a standard gross weight of 125,000 lb., but Boeing also is offering two optional higher weights of either 130,000 or 135,000 lb. Two zero fuel weight options—105,000 or 106,500 lb.—are being offered in conjunction with this. In comparison, the maximum gross weight of the 737-200 is 116,000 lb. in its standard configuration, or 125,000 lb. in a high gross weight configuration.

Fuel capacity of the newer aircraft is 5,311 U.S. gal. in the standard configuration, but Boeing is offering optional 370- or 810-gal. tanks that could be installed in the aft cargo compartment just behind the wing to provide an additional range of 200 naut. mi. or 400 naut. mi., respectively.

I programed the 737-300 flight management system for our flight, using the control/display unit in front of the throttles on the center console. Standard configuration of the aircraft provides only one CDU, but Boeing offers a second unit as an option. That unit is placed beside the first on the center console, giving both the pilot and copilot the possibility of communicating with the flight management computer.

I entered our route in the flight management system. The route was a fairly direct one—from Boeing Field to the Ellensburg VOR about 80 naut. mi. to the east, then northeastward another 47 mi. to the Moses Lake VOR, which is colocated at Moses Lake/Grant County Airport in eastern Washington.

An extensive navigation data base is stored in the flight management computer and contains all of the information needed by airline crews to plan and execute a routine commercial flight. Included are data on very high frequency (VHF) and low frequency navigation

Dual flight management system control/display units (CDUs) are installed on the third 737-300 flight test aircraft flown on an Aviation Week & Space Technology *demonstration flight. The units are shown here with the weather radar screen between the two units. Standard configuration provides only one unit on the captain's side, but most airline customers are opting for the second unit for the first officer as well.*

A closeup of the screen of one of the CDUs.

aids, en route waypoints, airways, airports, runways, standard instrument departures (SIDs), standard terminal arrival routes (Stars) and other data required to operate in a specific geographical area.

Boeing has made arrangements with Lear Siegler and Jeppesen Sanderson to update the navigation data base every 28 days on the 737-300, using magnetic tape cartridges and a portable data base loader. The updated data base can be loaded from the cockpit in about 10 min.

Flight crews also can enter waypoints and other information not in the stored data base. In this case, the route from Seattle to Moses Lake is one used frequently by Boeing flight test crews and was already stored in the navigation data base. I had only to select it using the flight management computer keyboard. Coordinates for our starting position had been entered into the system earlier, using the same keyboard.

The flight management computer also holds a performance data base containing the aerodynamic parameters of the 737 as well as engine information such as fuel flow, thrust and operating limits. Information on the amount of fuel on board the aircraft is fed directly to the computer from the fuel quantity sensor system. After entering the starting point and the route into the computer, the crew has only to enter the aircraft gross weight, cruise altitude and cost index—an indicator of the aircraft's operating costs—to allow the computer to calculate the most economical speed and altitude for the trip.

For example, when I entered our selected cruise altitude of 21,000 ft. (flight level 210), the flight management system indicated on its display screen that the most economical cruise altitude for our route would be 27,300 (flight level 273).

I also indicated our intention to perform an instrument landing system (ILS) approach and landing to Runway 32 at Moses Lake, so that the computer could complete its estimates of time and fuel requirements for the trip.

The CFM International CFM56-3 engines started relatively quickly. Elapsed time from the moment I pushed the engine start switch until the starters automatically deactivated themselves was about 40 sec. for each of the two engines.

Forward nacelles

The engines are housed in nacelles that extend forward from and are close-coupled to the wing of the 737-300, rather than hung on pylons. Despite the greater diameter of the General Electric/Snecma engines compared with the Pratt & Whitney JT8D-17 engines powering the 737-200, ground clearance is only slightly reduced. The CFM56 nacelles clear the ground by 18 in. at their lowest point, compared with 20 in. for the JT8Ds.

Program officials here said there has been some concern expressed by operators over the potential for foreign object damage during ground operations with the larger CFM56 engines. However, they do not expect the 737-300 to have any more of a problem with this than the 737-200, which they contend has the lowest damage rate of any JT8D-powered aircraft.

The 737's engines are outside the spray patterns for the aircraft's landing gear and thus are not prone to ingestion of objects thrown up by the tires, they said. In addition, the inlet velocities of the CFM56 engines are lower than those of the JT8Ds, they said, because of the higher bypass ratio of the CFM56.

Operators of Cammacorp/McDonnell Douglas DC-8 Super 70s powered by CFM56-2 engines encountered a high engine component erosion rate from dust and sand ingestion when those aircraft first entered service a few years ago, but 737-300 program managers here said they believe the fixes that were incorporated in the engines since then have made it extremely tolerant to dust and sand. There were no limitations placed on the ground operation of the CFM engines on this particular flight.

Gross weight of the aircraft was still around 100,000 lb. as we taxied onto Runway 13R for our takeoff from Boeing Field. The air temperature was 60F, and the tower was reporting winds from 100 deg. at 9 mph. I pushed the throttles forward to start the engines spooling up and depressed one of the takeoff/go-around buttons on the throttle levers to activate the automatic throttle system. The throttles then moved forward by themselves to the precomputed takeoff thrust level.

Quick acceleration

The relatively light weight of the aircraft combined with the high thrust of the CFM56 engines resulted in a quick acceleration to our rotation speed of 124 kt., and the aircraft was airborne after a roll of about 2,500 ft. Best single-engine climb speed (V_2) for the aircraft at that point was 133 kt., according to the flight management system computer, and the flight director provided the attitude guidance to achieve that speed plus 20 kt. immediately after takeoff. I raised the nose of the aircraft to follow the flight director commands.

McRoberts raised the landing gear and flaps, and then pulled the throttles back to 91% N_1 (engine fan speed) to reduce the thrust level somewhat. The flight director was still commanding a nose-up attitude of more than 20 deg. to keep our airspeed from exceeding the FAA limit of 250 kt. below 10,000 ft.

Our high climb rate was due to the uncommonly high thrust-to-weight ratio provided by the CFM56 engines. Even at its maximum gross takeoff weight of 125,000 lb. in the standard configuration, the 737-300 will have a thrust-to-weight ratio of 0.320 with both engines operating at their maximum thrust level of 20,000 lb. This compares with a thrust-to-weight ratio of 0.284 for the 737-200 at its normal maximum gross takeoff weight of 116,000 lb. with its two JT8D-17 engines generating 32,000-lb. thrust.

At its optional higher gross weights of 130,000 and 135,000 lb., the 737-300 will still have a substantially higher thrust-to-weight ratio than the standard 737-200. Furthermore, program officials at Boeing anticipate the availability of higher thrust versions of the CFM56 engine that will provide even higher performance.

Next year, for example, Pakistan Airlines is scheduled to begin receiving 737-300s powered by CFM56-3B2 engines producing 22,000-lb. thrust each. The airline has requested the higher thrust engines for better performance margins at the many high altitude/hot temperature airfields it serves.

Additional engine growth to 23,500-lb. thrust is achievable in the foreseeable future, according to James A. Von Der Linn, marketing manager for the Boeing Renton Div.'s 707/727/737/757 programs. He said further growth was envisioned beyond this, with accompanying improvements in specific fuel consumption.

We had been cleared initially by Seattle approach control to 13,000 ft. on our climbout from Boeing Field, but as we approached that altitude we were cleared to our requested cruise altitude of 21,000 ft. McRoberts called up a separate page in the flight management system computer showing the single-engine climb capabilities of the 737-300. At its present weight, the aircraft would have been capable of climbing with one engine inoperative to 22,700 ft., according to the computer.

Seattle approach control had vectored us onto our course almost immediately after takeoff, and after being cleared to 21,000 ft., I engaged one of the two independent channels of the Sperry SP300 autopilot-flight director.

Aural warning

An aural warning sounded briefly in the cockpit as we reached 20,000 ft. to announce that we were 1,000 ft. below our selected altitude. Our indicated airspeed was 273 kt. (Mach 0.61) as we passed through 20,000 ft. with the engines still set at 91% N_1 and the aircraft climbing at a rate of 2,200 fpm.

The autopilot smoothly leveled the aircraft at 21,000 ft. and a cruise speed of Mach 0.74, and the flight management system reduced engine power to 73.3% N_1 through the aircraft's automatic throttle system.

Incorporation of a 4% chord extension in the leading edge of the 737-300 wing compared with the 737-200 has resulted in a 4% increase in aerodynamic efficiency in the newer aircraft and a subsequent improvement of more than Mach 0.02 in the optimum cruise speed of the 737-300. It also has meant a reduction in fuel consumption of 1.5-3.0% at ranges from 200-1,500 naut. mi., according to program managers.

With the autopilot engaged and coupled to the flight management system computer, the aircraft followed our flight plan route as it was entered in the computer. When it was necessary to deviate from the planned route to comply with air traffic control instructions, such deviations could be commanded on the glare shield mode control panel without having to disengage the autopilot.

Heading vectors are the most common deviations, and it is possible to accomplish these on the 737 system by turning the "Heading Change" knob to the assigned heading. Altitude changes are accomplished in the same manner.

When air traffic control requirements took the aircraft off its flight plan route, a dedicated pointer on the horizontal situation indicator continued to point to the next waypoint on the flight plan route, giving a positive indication of which direction to turn once normal navigation was resumed.

As our flight progressed, the flight management computer continuously searched the navigation data base to select and automatically tune the best pair of VOR/DME (distance measuring equipment) stations from which to determine the aircraft's position.

When this automatic tuning process was under way, a white bar showed across the frequency readouts in the VHF navigation receivers to indicate they were being tuned selectively by the flight management system. The flight management computer then used the radio data to update the inertial reference system (IRS) position and compute an accurate aircraft position.

Vertical navigation

In addition to these lateral navigation functions, the flight management system also displayed its vertical navigation capabilities during this flight. As we leveled off at 21,000 ft., the system was already calculating the point at which to begin the most economical descent (engines at flight idle) for our arrival over Moses Lake. The "top of descent" point was displayed in the flight management system CRT display in terms of miles to go and time to go. The point was located about 3 mi. before the Ellensburg VOR.

Approximately 5 mi. before the indicated top-of-descent point, a message appeared on the flight management system control/display unit to select the next altitude. We were cleared to 11,000 ft., and I dialed in the new altitude in the altitude select window of the glare shield mode control panel.

Just before the aircraft reached the Ellensburg VOR, the throttles retracted automatically to the flight idle position, and the nose pitched slowly downward as the flight management system began the descent. During the descent, Grant County approach control cleared us to 7,000 ft. and direct to Pelly, a low-frequency beacon that serves as both an initial approach fix and the outer marker for the Moses Lake ILS approach to Runway 32R. I selected 7,000 ft. on the autopilot-flight director mode control panel while McRoberts entered Pelly on the "direct to" line of his flight management system control/display unit and pushed the "execute" button.

This ability to reprogram the flight management system quickly meant that neither McRoberts nor I had to touch the controls of the aircraft—something we had not done since engaging the autopilot during our climbout from Seattle.

We crossed Pelly at 4,000 ft. and continued descending to 3,000 ft. as I dialed in a right turn with the heading selector on the mode control panel to take the aircraft outbound a few miles before turning farther to intercept and track inbound on the ILS localizer.

Turning the heading selector farther right to steer the aircraft to the localizer, I pushed the "Approach" button on the mode control panel and armed the second autopilot channel. This engaged the Category 3 automatic landing system that is part of the 737-300 flight management system options package and relieved me of the need to make any further heading or altitude inputs. The aircraft flew through the ILS localizer on its first attempt to intercept the beam because our outbound track had been too close to the inbound course, leaving the aircraft too little room to make its inbound turn.

The system recovered nicely from the overshoot, however, and intercepted the localizer beam successfully from the other side. The system also maintained our 3,000-ft. altitude until the aircraft intercepted the ILS glideslope about 5 mi. from the runway.

McRoberts lowered the landing gear and extended the flaps incrementally as the aircraft approached the glideslope. I observed but did not have to touch the control yoke, which moved vigorously fore and aft, making indiscernible attitude changes needed to compensate for the moderate turbulence and keep the aircraft on glideslope.

At 1,500 ft. above the ground, the automatic landing system went through a final series of self-tests before advancing to the landing mode. The rest of the approach and landing was uneventful, with the aircraft touching down just slightly to the right of centerline. I disconnected the autopilot and took off manually to prepare for another approach.

There is little difference in the manual handling qualities of the 737-300 and the 737-200, with the newer aircraft showing the same appreciable responsiveness and agility as its predecessor.

As the aircraft crossed the runway threshold on our second automatic approach, the takeoff/go-around switch on one of the throttle levers was depressed, and the aircraft executed an automatic go-around.

Protective features

On the downwind leg of our third approach, McRoberts demonstrated some of the protective features built into the automatic throttle system. With the autopilot engaged and the aircraft at a speed of about 200 kt. and an altitude of 2,000 ft. above the ground, McRoberts suggested that I select a speed on the mode control panel that was below that required to maintain flight. I selected 100 kt.

Our flaps were completely retracted, and as the aircraft's speed dropped below 140 kt., the throttles moved forward automatically to maintain a safe flying speed. Throttle movements under these circumstances were related to angle of attack rather than airspeed, and the autothrottle system is programed to maintain a safe angle of attack no matter what the configuration of the aircraft.

McRoberts then extended full landing flaps and suggested I accelerate the aircraft. Although I applied nearly full power initially, the automatic throttle system reduced thrust as the aircraft approached the not-to-exceed speed of the aircraft with flaps fully extended.

I flew a manual Flight Director approach and landing on our third approach to Moses Lake, using the autothrottle system to maintain airspeed stability in the gusty, thermal conditions we were encountering. The aircraft was easy to fly under these circumstances, and the approach and landing were normal.

As I applied power and rotated the nose for takeoff after that landing, McRoberts brought the right throttle to idle, simulating an engine failure on takeoff. The onset of yaw and roll to the right because of asymmetric thrust was very slow in developing, allowing plenty of time to counter this with additional left rudder and left rudder trim.

I made a 90-deg. turn to the left followed by a 270-deg. turn to the right to establish the aircraft on a final approach to Runway 14, still with one engine inoperative. The aircraft handled comfortably in this pattern with a speed of 140 kt., and I slowed to 130 kt. on final approach. After a normal touchdown, McRoberts brought the right engine on line again, and I made a normal two-engine takeoff to return to Boeing Field.

We climbed to 16,000 ft. for the return flight, and I slowed the aircraft to observe the automatic slats on the leading edge of the 737-

300. The slats could be seen extending by looking aft from the side windows of the cockpit as the aircraft slowed. There was no sensation of slat movement inside the aircraft, however.

Still at 16,000 ft., I slowed the aircraft to the onset of a stall with the autopilot and autothrottles disengaged. The aircraft weight was approximately 92,000 lb. at that point, and wing buffeting became discernible as airspeed declined to approximately 105 kt.

Full stall

At McRoberts' suggestion, I continued pulling aft on the control yoke until it reached its aft limit. With the intensity of the buffeting still increasing, the aircraft continued slowing until it reached a full stall at 95 kt. indicated airspeed. The wings remained level and the nose remained slightly above the horizon as long as I held the yoke aft. The aircraft gave the impression of remaining completely controllable in the pitch, roll and yaw axes. I pushed the yoke forward, and the aircraft immediately pitched over and began to accelerate out of the stall.

I applied full power at that point, allowing the aircraft to accelerate to its maximum speed. Maximum operating speed (V_{mo}) for the aircraft at our 16,000-ft. altitude under the prevailing conditions was 340 kt., and as the 737-300 approached that speed, the automatic throttle system began backing off the throttles to keep the aircraft from exceeding the limit. Like the angle of attack floor at the low-speed end of the flight envelope, the overspeed protection feature is active even when the autothrottles are not engaged.

An uneventful automatic approach to Boeing Field followed. We landed after 1.7 hr. of flight with 14,400 lb. of fuel remaining out of the 24,150 lb. in the aircraft at engine start.

Because of the high degree of commonality between the 737-300 and the 737-200, project engineers and pilots here believe that airline crews already qualified on 737-200s equipped with automatic flight control systems will be able to qualify in the 737-300 with only two or three days' training in the differences between the two versions. This training would concentrate mainly on the engines, the flight management system and variations in hydraulic systems.

Crews qualified in 737s not equipped with automatic flight control systems are expected to need only about five days' training, with the extra two days dedicated to the functions of the automatic flight control system. Boeing has proposed to the FAA that the crews in these two categories should be qualified, once they complete the training programs, to operate the 737-300 without the need for an additional type rating. Crews not currently qualified in the 737 would be put through a 26-day course.

Chapter 3

Regional

Regional turboprop for the 1990s

David M. North/Prestwick, Scotland
December 14/21, 1992

Low acquisition and operating costs are pluses for the Jetstream 41, the newest entrant in its 30-seat class.

British Aerospace is attempting to compensate for its late entry into the 30-passenger regional aircraft market by offering the twin turboprop Jetstream 41 as an entry level aircraft with a relatively low acquisition and operating cost.

The British Aerospace turboprop aircraft subsidiary has the reputation of building yesterday's aircraft today. This is primarily founded on the two other commercial turboprops built by the British company. The larger ATP's heritage can be traced to the Hawker Siddeley 748, which first flew in the early 1960s. The ATP is now being built here as part of a consolidation program.

More closely allied with the Jetstream 41 is the Jetstream 31 series of regional aircraft, the 19-passenger commuter aircraft dates back to the Handley Page Jetstream developed in the mid-1960s and equipped with two Societe Turbomeca Astazou turboprop engines.

In developing the Jetstream 41, British Aerospace combined its past experience with current technology to build an aircraft that would find a niche in an already crowded market. The company lagged in introducing a regional aircraft of this class. The Jetstream 41's closest competitor, the Embraer EMB-120 Brasilia, started flying in revenue service

British Aerospace's Regional Aircraft division plans to deliver three twin-turboprop Jetstream 41s this year, 33 next year and 45 in 1994.

in 1985. The Saab SF340, another close competitor, entered service in 1984. The de Havilland DHC-8 entered service the same year. All three aircraft companies have upgraded their earlier 30-passenger aircraft.

A realistic British Aerospace assessment of the regional aircraft market showed the Jetstream 41 should carry no more than 29 seats and that it had to be priced below the competition and have lower operating costs. The price of the well-equipped Jetstream 41 in its standard configuration is about $6.7 million. The company also claims operating costs more than 10% less than its closest competitors.

The quest to build a 29-passenger entry level regional aircraft did not allow British Aerospace to push for advanced technology in the Jetstream 41. The Jetstream 41 flown here lost month, however, easily qualifies as today's aircraft being built today.

While the Jetstream 41 externally resembles the smaller Jetstream 31, the larger aircraft is much more advanced in systems, avionics and manufacturing process. The difference is so significant that the Jetstream 41 has received a separate type certificate from the European Joint Aviation Authorities, and similar treatment is expected from the U.S. Federal Aviation Administration early next year.

The Jetstream 41 flown here also is much different in performance and handling qualities than the Jetstream 311 flew in late 1982 (AW&ST Jan. 3, 1983, p. 43). The Jetstream 41 has more precise handling characteristics than the smaller aircraft and feels like a larger aircraft than you would expect from its 23,000-lb. maximum takeoff weight. This often is the case when an aircraft's fuselage is stretched.

Manx Airlines, based on the Isle of Man, and Loganair of Glasgow, Scotland, received their first 29-passenger Jetstream 41s in late November. U.S. FAA Part 25 certification of the Jetstream 41 is expected in late February.

The handling qualities of the Jetstream 41 got a strong test during the evaluation flight in Western Scotland's adverse late November weather. The evaluation flight was with Tim N. Allen, manager of flight test for the Regional Aircraft division here. Following a pilot briefing in the operations office, Allen detailed some of the major differences between the Jetstream 31 and 41 during the preflight inspection. The most apparent external difference is the fuselage stretch. A 8.2-ft. (2.5-meter) plug in front of the wing allowed the company to install a forward passenger door with airstairs. A 7.8-ft. (2.3-meter) plug also was added behind the wing.

The main landing gear now retracts forward into the nacelles designed for the Garrett TPE33 1-14 engines. The longer wing span and the five-bladed McCauley propellers allowed the propeller tips to be placed farther from the fuselage. The larger wing also afforded an increased fuel capacity. The flaps were redesigned and are more effective with an increased span and chord. In addition, spoilers were installed inboard of the nacelles.

I occupied the left seat while Allen took the right seat. The Jetstream 41 has floor mounted control columns, and entering the seat took a little more care than with other aircraft of this class. The large center console did not help.

The Jetstream 41 flown, registered as G-JMAC, is the fourth one built. It was the first of the test Jetstream 41s with a complete interior. Gross weight at the ramp was 20,200 lb. (9,163 kg.) or 2,910 lb. (1,320 kg.) short of its maximum 23,110 lb. (10,483 kg.) ramp weight. The fuel at the blocks was 5,800 lb. (2,630 kg.) and within 200 lb. (90 kg.) of the aircraft's maximum usable fuel.

Loganair Jetstream 41 flies near bridge spanning the Firth of Forth in Scotland. About 40 months passed between start of design and certification.

Allen started the right engine with the overhead start switch using external power. I started the left engine with the some automatic procedure, and engine exhaust temperature stayed well below maximum on the 50F (10C) day. The engine indications are shown digitally and by an arc presentation on the Smiths engine management display.

The V_1 takeoff decision speed and V_r rotation speed was calculated by Allen to be 104 kt. The V_2 takeoff safety speed was 111 kt. Allen said the V_1 speed will be separated from V_r after more testing to gather better overall takeoff performance data. The takeoff reference speed was displayed on the lower left side of the Honeywell Electronic Attitude Direction Indicator (EADI).

I taxied the Jetstream 41 to Runway 13 at the Prestwick airport. The wind was calm in light rain showers. The nosewheel steering handle was used effectively throughout the taxi with only a short time required to gain confidence in the system. With the condition levers in the taxi position, I found that I did not have to move the power levers often to maintain a comfortable speed. The brakes were effective and did not have a tendency to grab during taxi. The flap setting was 9 deg. for takeoff. The fuel used during the 16-min. engine start to takeoff was 110 lb.

The Jetstream 4l was aligned on the runway, and I held the brakes for a standing takeoff while advancing the power levers. Allen aligned the throttles to a 102% torque setting, and I released the brakes. Allen held the yoke forward while I used the nosewheel steering for directional control until reaching 70 kt. At that point, rudder was used for directional control and I took the yoke. After a ground roll of 2,200 ft., the nose was rotated to a pitch attitude of 8 deg. The initial force required to reach the takeoff pitch attitude was higher than usual. Allen had warned me that with its forward center of gravity this was to be expected.

British Aerospace conducted its hot and high altitude testing with the No. 2 Jetstream 41 prototype flying from Roswell, N.M., last summer.

The Jetstream's landing gear and flaps were retracted and a 170-kt. climb was initiated while turning to a westerly heading. Because of the weather, the aircraft's anti-icing had to be used during the latter portion of the climb.

Passing through 5,000 ft. after takeoff from the sea level airport required less than 3 min. At that altitude the rate of climb was 2,600 fpm. The rate of climb dropped to 1,600 fpm passing through 10,000 ft. and total fuel burned was 248 lb. Slightly less than 9 min. was required to reach 15,000 ft., and the rate of climb was 1,000 fpm. Total fuel used at that point was 315 lb.

The assigned flight level of 20,000 ft. was reached in 13.5 min. with a total fuel burn of 408 lb. The aircraft was in clouds during the entire climb, and I was able to observe the ice being shed from the outer left wing at the higher altitudes. The temperature at 20,000 ft. was standard conditions plus 6°C. With the anti-icing system on, a 63% torque setting allowed a 201 kt. indicated speed and a 920 lb./hr. fuel burn. The resultant true airspeed with engine bleed air being used for deicing was 274 kt.

Allen turned off the anti-icing system to obtain maximum cruise figures in more normal conditions. The torque increased to 69% and the total fuel flow to 968 lb./hr. The indicated airspeed was 210 kt. and the true airspeed 286 kt. at 20,000 ft. The regional aircraft officials estimate they are shy 3 kt. of their 295 kt. maximum true airspeed goal, but that minor modifications will make up the difference.

With the anti-ice system still off, a 51% torque indication was set for long-range cruise results. The indicated airspeed was 185 kt. and true airspeed was 254 kt. The total fuel flow at this speed was 776 lb./hr.

Allen decided that we should try the Inverness area to find some clear sky to perform handling and stall maneuvers. During the cruise to the more northern area, I had the opportunity to better observe the Jetstream's cockpit layout and the performance of the autopilot. The Regional Aircraft officials have designed the cockpit with a mixture of electronic indicators and round dial instruments. Instead of placing the altimeter in the EADI, like many other displays, they had Honeywell place the rate of climb indicator on the right side and a separate altimeter was installed to the right of the EADI. I found that this arrangement was easy to assimilate into my cockpit scan. The human engineering needed for effective situational awareness was evident in the design.

The Jetstream 41 is equipped with an integrated radio management unit on the front instrument panel. Allen said this unit was offered by Honeywell and was one of the reasons that the company was chosen to provide the four-tube electronic displays. There are no

Jetstream 41's standard interior includes 29 seats with a 30-in. pitch, a rear lavatory and a rear galley. Main baggage compartment is behind the cabin.

plans to integrate check lists or system symbology on the electronic displays.

One throwback to the Jetstream 31 is placing system controls and gauges on the front center console. For example, the hydraulic controls, a low use system during flight, would be better placed elsewhere. In its place, an integrated area navigation system could be installed, rather than on the front instrument panel as planned now.

Placing the Honeywell SPZ-4500 autopilot control panel in the rear of the center console was an excellent choice. The easy-to-use autopilot's functions all were smoothly executed with no surprises. The Jetstream 41 also can be equipped with a Primus 650 color radar with the presentation on the horizontal situation display.

After finding relatively clear air near Inverness, I used the autopilot to descend at 1,500 fpm to 10,000 ft. The descent was at maximum operating speed (V_{mo}), and the noise level in the cockpit was not excessive.

Steep turns at 180 kt. were flown up to bank angles of 60 deg. Slight power additions were required to maintain speed, but the aircraft was stable. During the maneuvers, I found that pitch control was very positive, but at the higher speeds the roll rate force was slightly higher than I would have preferred.

The aircraft was slowed with the engine rpm at 100% and the torque set at less than 30%. I trimmed the aircraft to 120 kt. and then continued to decelerate at a rate of 1 kt./sec. The aircraft's stick shaker vibrated at 105 kt. in the clean configuration, and the stick pusher activated at 98 kt. In the landing configuration, the stick shaker came on at 89 kt., and the stick pusher activated at 84 kt. Altitude loss was in the neighborhood of 200 ft. in the clean configuration and 400 ft. in the landing configuration.

The aircraft's flight controls were responsive all the way into the stall regime, and recovery was straightforward. The stalls were accomplished with the anti-ice system off. Had the system been on, the angle-of-attack limit for the stick shaker and pusher would automatically have been lowered and their activation speeds would have been higher.

Allen then shut down the left engine at 9,000 ft., and I was able to confirm that human engineering needed for effective situational awareness was evident the propeller was feathered. The Jetstream 41

Honeywell SPZ-4500 avionics system, including a four-tube electronic display, is standard.

has counter rotating propellers, helping to offset the yaw with an operating right engine. Once the yaw was zeroed out by use of rudder trim, the aircraft was stable and there was ample rudder still available. With an 88% torque setting, the indicated airspeed at 9,000 ft. was 247 kt. and the true airspeed was 285 kt. Fuel flow on the one engine was 625 lb./hr. The left engine restarted without difficulty, and I descended for two touch-and-go landings at the Inverness airport.

The selection of 9-deg. flaps at 160 kt. did give a slight ballooning effect. The landing gear was lowered and 25-deg. flaps selected for the initial straight-in visual approach. I flew the final approach at V_{ref} plus 10 kt., or at 118 kt. The touchdown was firm at a landing speed of close to 105 kt. I had not allowed for enough flare on the first landing. A short landing pattern was flown for the second attempt, but with more of a flare the touchdown was smoother.

The only fault in the aircraft during landing was that I would have preferred the engine be more responsive to throttle movements at the low power end, especially near touchdown. During the return trip to Prestwick, Allen flew while I went back to the cabin. The noise level directly in line with the propellers was quite high, while in the rear of the cabin it was comfortable. The Regional Aircraft officials are working to lower the noise level in the area where the propeller wash impinges on the top of the fuselage.

British Aerospace is building the Jetstream 41 in 35 days on its reconfigured, more efficient line.

The trip back to Prestwick become more turbulent in heavy rain as we approached the airport. The Jetstream 41 exhibited excellent stability and had the feel of a much larger aircraft during the turbulence.

The landing at Prestwick in heavy rain was smoother than the other two. The cockpit presentation for the modified instrument landing approach was easy to follow, and I had no difficulty with instrument scan. The Jetstream 41's cockpit instrumentation has been designed for commuter operations in congested airports and all types of weather.

The Jetstream 41 was taxied back to the ramp at the Regional Aircraft facility. The total flight time was 1.8 hr. with three landings. The block-to-block time was 2.2 hr., with a total fuel burn of 1,900 lb.

Nick Godwin, the Regional Aircraft vice president of marketing services, believes the Jetstream 41 will replace 19-seat commuters as the standard entry level aircraft for regional operators. This belief is based on the benefits of certification to Part 25 FAA requirements, operational and passenger flexibility and support standards.

British Aerospace holds 25 firm orders for the Jetstream 41. Two of the twin-turboprops were delivered late in November, one to Loganair and one to Manx Airlines. Loganair holds orders for an additional two aircraft, while Manx has orders for four more Jetstream 41s. Fifteen Jetstream 41s are on order by Atlantic Coast Airlines, with the first delivery scheduled for March, 1993. Sun-Air of Scandinavia has orders for two aircraft.

JETSTREAM 41 SPECIFICATIONS

POWERPLANT

Two Garrett TPE331-14 GR/HR turboprop engines flat rated to 1,500 shp. Thermodynamic rating is 1,760 shp. An automatic power reserve extends the engine rating to 1,946 shp. after an engine failure at takeoff.

PROPELLERS

Two McCauley five blade, 114-in-dia., constant speed with full reverse and feather capabilities.

WEIGHTS

Maximum ramp weight	23,110 lb. (10,483 kg.)
Maximum takeoff weight	23,000 lb. (10,433 kg.)
Maximum landing weight	22,300 lb. (10,115 kg.)
Maximum zero fuel weight	20,700 lb. (9,390 kg.)
Typical operating weight empty	14,000 lb. (6,350 kg.)
Maximum payload	6,700 lb. (3,040 kg.)
Maximum usable fuel	5,960 lb. (2,703 kg.)

DIMENSIONS

Length	63.2 ft. (19.3 meters)
Height	18.8 ft. (5.7 meters)
Wingspan	60 ft. (18.3 meters)
Cabin height	5.8 ft. (1.8 meters)
Cabin width	6.1 ft. (1.86 meters)
Cabin length	31.3 ft. (9.54 meters)
Checked baggaged vol.	217.5 cu. ft. (6.16 cu. meters)
Standard capacity	29 passengers

PERFORMANCE

Range with 29 passengers at standard conditions	670 naut. mi.
Range with a 3,000-lb. payload at standard conditions	1,600 naut. mi.
Takeoff requirement at 23,000 lb., standard conditions	5,000 ft.
Landing requirement at 22,300 lb., standard conditions	3,930 ft.
Maximum cruise speed	292 kt.
Maximum certificated altitude	26,000 ft.
Maximum cabin differential	5.7 psi.

Fokker 50 combines handling ease with quiet operations

Bruce D. Nordwall/Amsterdam
July 18, 1988

The Fokker 50 is a quiet, efficient turboprop, with excellent handling qualities for the midrange, 50-passenger commuter market. Fokker chose the middle ground in the cockpit design, offering the flexibility of an electronic flight instrument system (EFIS) but the cost savings of analog engine instruments.

The aircraft received its certification from the RLD, the Netherlands Dept. of Civil Aviation, on May 15, 1987. The first production aircraft was accepted by the West German regional airline DLT in July, 1987 (AW&ST Feb. 15, p. 110).

This *Aviation Week & Space Technology* pilot flew an evaluation flight in the No. 2 prototype from the Fokker facility at Schiphol Airport, Amsterdam. I also had the opportunity to ride along during an Ansett Airlines acceptance flight and observe the comfort and accommodations of the 13th production aircraft.

Jaap Hofstra, Fokker 50 project pilot, conducted the preflight briefing and flew in the right seat during the evaluation. Flight engineer Frans Ketting assisted and recorded data from the jump seat.

The Pratt & Whitney PW125B engine is based on the well-known FT6, but with greatly improved efficiency, Hofstra said. It is a three-

Fokker 50 aircraft destined for Scandinavian Airlines System is airborne over the North Sea on a check flight before deliver. The Fokker 50 was certified in May, 1987. Fokker has 84 firm orders and 26 options for the aircraft, including 15 orders and 15 options from SAS.

shaft engine with a two-stage, free-power turbine driving the propeller through a reduction gearbox.

An improved propeller synchronization system reduces harmonics between the two engines as well as noise level inside and outside the aircraft. Propeller rpm is also relatively low, controlled between 1,200 rpm at takeoff and 1,020 rpm during climb and cruise. The control mechanism used for propeller synchronization is an adaptation of a stability augmentation system for a fighter aircraft flight control system, Ketting said. There were no propeller problems during the test program, he said.

The six-blade propellers of the Pratt & Whitney 125B engines are the most obvious visual distinction from the F-27. Dowty Rotol designed the propellers specifically for the Fokker 50. The lower tip-speed contributes to quiet operation. Propeller performance during takeoff and climb is increased by the use of composite construction blades with a high aerodynamic efficiency.

Two basic configurations of the Fokker 50 are available—the standard 41,865-lb. (18,990-kg.) maximum takeoff weight aircraft with a range of 584 naut. mi., and one with an extended range capable of 1,467 naut. mi., uprated to 45,900-lb. (20,820-kg.) maximum takeoff weight. Although the lower weight version was intended to let European operators benefit from a price break in operating fees, all sales have been for the upgraded version.

During normal operation the pilot controls the engines and propellers by pushbutton selections on the engine rating panel, which sends inputs to the engine electronic control (EEC) and propeller electronic control (PEC). The pilot places the power levers in a detent and then selects power desired—takeoff, go-around, climb, cruise, maximum continuous thrust or a flexible setting that the customer can have preset.

The EEC calculates and regulates the engine torque for the selected power and ambient conditions and commands the applicable propeller rpm. Reducing from takeoff to climb power requires just pushing one button, reducing cockpit workload at a critical flight phase.

The propeller control has been simplified, compared with previous generation of propeller controls, to increase reliability. The propeller is hydraulically actuated under a dual-channel electronic control. Only one channel is in operation, but the system automatically switches to the backup channel in case of a failure.

Ketting started the preflight while Hofstra and I checked the weather, which was forecast to be VFR for the duration of our flight. On preflight, both the six-blade propeller and the shielded, swept-up aileron horn balance at each wingtip—called a Folklet by the company—got my attention. The Folklets improve lateral/directional stability and pitch/roll control harmony, Hofstra said.

To facilitate quick turnaround, the aircraft has four doors, one on each side forward and aft, which allow simultaneous passenger, baggage, servicing and catering access. The Fokker 50 has single-point and overwing fueling. We boarded the aircraft and I took the left seat in the comfortable cockpit. The dual electronic flight instrumentation system has a color CRT for primary flight display and navigation display for both pilots.

The engine instrument panel consists primarily of conventional gauges, arranged from top to bottom, with the instruments that are most important to the pilot on top. A small window with a digital readout also has been added to the gauges for propeller rpm, high-pressure turbine rpm and interturbine temperature (ITT).

Four doors on the Fokker 50 contribute to a rapid turnaround. Separate entrances are provided for passenger loading, servicing, cargo and luggage.

I found the combination comfortable, being able to scan the needles quickly and to get precise digital readings when needed. But the primary reason for installing those analog instruments was cost effectiveness, according to Fokker. The notable exception to the round gauges is the vertical tapes for torque, the primary engine control parameter. If the pilot has selected an engine rating with a push button on the engine rating panel, the EEC will compute the target torque and display it with a pointer (bug) on the torque indicator so the pilot can verify performance.

The Fokker 50 has an audio control panel with toggle type switches on slide potentiometers to control the volumes on the three VHF and two HF radios. I found the slide potentiometers particularly easy to use by touch alone.

Hofstra depressed the master engine start button in the overhead console, and rotated the spring-loaded start switch to start the right engine. In both cases we had oil pressure indication within 3 sec., the ITT indicated about 600C, well below the starting limit of 950C, and high-pressure turbine speed stabilized at 74% as expected. I then started the left engine. We started the engines using external a.c. power, but could have used external d.c. power or the two 28-v. aircraft d.c. batteries. An auxiliary power unit will be ready as an option next year.

Fuel control is automatic. Before taxi, the pilot sets the aircraft weight from the load sheet into the aircraft weight and fuel flow panel. The system subtracts fuel used and keeps a running total of aircraft weight.

Integrated alerting system

The Fokker 50 integrated alerting system warns of abnormalities in systems or operations. The alerts conform to the international standards and range from level one to level three warnings, the latter requiring immediate corrective action. During the takeoff phase, only the most important engine alerts are shown to avoid unnecessary confusion and aborts.

We taxied to Runway 24, completing the short takeoff checklist en route. The weather was reported as clear with 10-kt. winds at 210 deg. Visibility was 10 km., temperature 5C (41F). Maximum takeoff weight is 45,900 lb. (20,820 kg.), but our ramp weight before start was 36,382 lb. (16,503 kg.) including 6,886 lb. (3,130 kg.) of fuel. Our center of gravity was 28.6% of the mean aerodynamic chord. Limits are 18.9 to 40.7%.

Ketting had computed critical engine failure speed, V_1, and rotation speed as 96 kt. Takeoff/climb speed with one engine inoperative, V_2, was 98 kt. We set the flaps to the normal takeoff setting of 5 deg. and set the rudder, aileron and elevator trim at 0 deg. We burned

44 lb. (20 kg.) fuel in the 6 min. from the time we started taxiing until takeoff.

On the runway I used the takeoff configuration warning system for one last check of proper configuration, then advanced power levers to the takeoff detent. The torque indicators showed 95%, the takeoff power for the ambient conditions. I rotated at 96 kt., cleaned up the aircraft during climbout and depressed the climb power push button on the engine rating panel. The EEC gave climb torque and the low propeller speed, 1,020 rpm, and we accelerated to 160 kt. for the most economical climb. As we climbed to the northwest, the aircraft felt stable and responsive to control inputs. The flight controls are completely mechanical, but the aircraft did not feel heavy.

Climbing out at 2,000 fpm, we dialed in our assigned altitude of 9,000 ft. on the flight mode panel and heard a warning bong passing 8,000 ft. At 9,000 ft., I selected cruise on the engine rating select panel and the EEC reduced power. Our stair-step climb under Schiphol departure radar control had taken 10 min. from the start of takeoff and consumed 198 lb. (90 kg.) of fuel.

Start of pressurization

The aircraft is pressurized to maintain cabin altitude at a maximum of 8,000 ft. at its maximum operating altitude of 25,000 ft., Hofstra said. With many aircraft, the start of pressurization causes a noticeable pop in one's ears as the weight-on-wheels switch senses the aircraft is airborne. The Fokker 50 starts pressurizing at the beginning of the takeoff roll, a method designed for more comfort as pressurization rises. With the beautiful day over the Netherlands, we had no need for the bad-weather systems, but I asked Ketting about the Fokker 50's capabilities. Bleed air is used to deice the leading edge of the wings and the horizontal and vertical stabilizers. Anti-ice protection is provided to the propeller blades, spinner and intake from the 40-kva. integrated-drive generators on each engine.

The aircraft is well outfitted in communication and navigation gear. Standard equipment includes dual VHF communication radios, dual VHF navigation with distance measuring equipment, and an ADF. Options include HF and UHF radios and VLF/Omega navigation.

Stalled characteristics

On reaching the operating area, I put the aircraft through a series of turns of increasing bank. The aircraft was solid, and I quickly acclimated to the mechanical elevator trim, only occasionally thumbing the yoke where an electrical trim would be. Production aircraft have an electrical aileron trim, which will be nice for the slight trim changes required with each change in power setting. During a timed turn I found one small feature on the yoke that pilots will appreciate during instrument flying—a button to control the elapsed time clock. Not having to reach to the instrument panel is a nice touch.

After a clearing turn, I reduced power, eased the nose up and let the aircraft decelerate into a wings-level stall. The stall warning stick shaker came on at 97 kt., and airframe buffet started at the wings level and at 90 kt. the nose dropped almost straight through. After recovering, I put the aircraft in a landing configuration with 25-deg. flaps and idle power, and held the aircraft nearly level as we decelerated. This time the stall warning activated at 78 kt. and the onset of buffet was at 77 kt. The buffet increased until the aircraft stalled at 75 kt. I had effective roll control down to the g-break, where an honest nose-down pitch made a straightforward recovery easy.

At 110 kt. and 10,500 ft., with the aircraft still in a landing configuration, I pulled the right throttle to idle. Hofstra feathered the engine and I cleaned it up. It did not take a leg-shaking amount of rudder to compensate for the asymmetrical thrust, and I was able to trim it all

Fokker designed the aircraft's cockpit as an affordable balance between electronic instruments and conventional round gauge instruments.

Fokker paints customer colors on aircraft before the outer wings are attached on the production line. Aircraft for four of the eleven airlines acquiring Fokker 50s are visible.

out. I left maximum continuous thrust on the left engine and held our altitude to see what speed could be attained. We accelerated to 183 kt.

Hofstra unfeathered the engine and brought it back on the line, and we returned to Schiphol for some landings. Approach control reported the weather as clear, greater than 10 km. visibility, with wind from 210 deg. at 16 kt., and identified the active runway as 19R.

Schiphol approach control provided radar vectors to intercept the ILS course inbound. We descended to 2,000 ft. and transitioned to a landing configuration with 25 deg. flaps and reduced to 102 kt. airspeed, which was V_{ref+10}, as we intercepted the glide slope at the 9-mi. DME fix. Hofstra recommended that I hold that airspeed during the approach and arrive over the runway threshold at V_{ref}, which I did. We had good ILS needles, and the aircraft trimmed well and held steady on the glide slope. On touchdown Hofstra reset the flaps to 5 deg., and zeroed the trim. The aircraft leaped smartly into the air when I rotated, and we turned downwind for another touch-and-go.

The landing gear, flaps, brakes and nosewheel steering are all actuated by the 3,000-psi. hydraulic system, driven by a pump on each engine. In the case of hydraulic failure, the flaps can be operated electrically, the landing gear free-falls, and the brakes can be operated using an accumulator. The antiskid brake system is always on.

I wanted to execute both normal and no-flap landings from the VFR pattern. The turboprop engine responded rapidly to power changes, and it was easy to sense these changes to help gauge the power needed for corrections in the pattern. The normal, 25-deg. flap landing was comfortable, so I set up with a slightly wider pattern for a no-flap touch-and-go. I slowed to the recommended no-flap airspeed, 115 kt. passing abeam, and turned base. Except for the higher airspeed and pitch attitude, there were no significant differences in aircraft feel. After the touch-and-go we turned downwind for a full-stop landing, using 102 kt. and 25-deg. flaps, as on the earlier full-flap touch-and-go.

Everything went well until rollout, when I lifted the thrust levers to pull the props into reverse. I had neglected to get the feel of this movement and failed to set them into reverse at precisely the same instant. The quick swerve was correctable, and anyone more used to activating thrust reversers would probably not have had any problem.

We taxiied in at ground idle thrust, which gave us a taxi speed of about 15 kt. At shut down, after the 2-hr. 10-min. flight time, we had burned 2,376 lb. (1,080 kg.) of fuel.

Ansett acceptance flight

After a debrief I had the opportunity to ride along while Ansett Capt. Keith Duncan flew the 13th production aircraft on an acceptance test flight for the Australian airline. The cabin, with 2-×-2 seating, was comfortable and had the feel of a larger aircraft. The standard configuration is 50 seats at 32-in. pitch. The Fokker 50 configuration can range from 46 passengers in business class with 34-in. pitch to high-density seating of 58 passengers at 30-in. pitch.

I thought the cockpit of the prototype I had flown earlier in the day was comfortably quiet, but was interested in the noise in the passenger cabin. For a worst case test, I sat in the window seat closest to one propeller, and I asked another passenger to take the corresponding seat across the aisle. When the pilot throttled back from takeoff to climb power, I found we could talk easily at slightly higher than normal speaking volume. At cruise power there was no difficulty in conversation.

Fokker has taken a lot of care to reduce noise, both for passengers and those on the ground. Triple-pane windows and a variety of sound absorption techniques were used in the cabin. Fokker claims to have the lowest interior noise of any commuter turboprop, and exterior noise well below ICAO and U.S. Federal Aviation Regulations.

Design of the aircraft relied on adhesive metal-to-metal bonding and composites. The benefits of this approach include weight saving, an aerodynamically smoother outer skin and corrosion resistance. All contribute to lower operating costs. Only 22 passengers per flight are needed to break even, according to Fokker, leaving 28 seats for profit.

Latest customers

Fokker has 84 firm orders plus 26 options for the Fokker 50. The largest orders are from Ansett (15 firm orders) and SAS (20 firm orders and options for 10). The most recent delivery was received by Maersk Air of Denmark, which will use the Fokker 50 on its Billund-London route. The latest customers are Austrian Airlines, which increased its order from two to four in March, and Scandinavian Airlines System. SAS ordered 15 and has options for 15 more.

FOKKER 50 SPECIFICATIONS

POWERPLANT

Two Pratt & Whitney Canada PW125B turboprop engines with a normal takeoff rating of 2,250 shp. An automatic power reserve feature enables each engine to be operated at its maximum 2,500-shp. rating with the failure of one engine.

PROPELLERS

Two Dowty Rotol six-blade, 12-ft.-dia., constant speed with full feather and full reversible capabilities.

WEIGHTS

Fokker offers the aircraft in a standard maximum takeoff weight of 41,865 lb. or an optional maximum takeoff weight of 45,900 lb. The difference is only in certification, not in structure. Data are shown only for the optional version since all orders are for that version.

Maximum takeoff weight	45,900 lb. (20,820 kg.)
Maximum landing weight	43,500 lb. (19,731 kg.)
Maximum zero fuel weight	41,000 lb. (18,597 kg.)
Operational empty weight	28,090 lb. (12,741 kg.)
Maximum payload	12,910 lb. (5,856 kg.)
Fuel capacity	9,090 lb. (4,123 kg.)

DIMENSIONS

Length	82.83 ft. (25.25 meters)
Wheelbase	31.82 ft. (9.70 meters)
Wingspan	95.15 ft. (29.00 meters)
Wing area	754 sq. ft. (70 sq. meters)
Wheeltrack	23.62 ft. (7.20 meters)

Cabin length	52.36 ft. (15.96 meters)
Cabin width	8.2 ft. (2.5 meters)
Cabin height	6.4 ft. (1.96 meters)
Standard cabin configuration	50 passengers at 32-in. pitch

PERFORMANCE

Takeoff field length requirement at 45,900 lb., and with flaps at 15 deg.	4,450 ft. (1,360 meters)
Landing field length requirement at 43,500 lb. and with flaps at 35 deg.	3,710 ft. (1,130 meters)
Maximum cruise true air speed at 16,000 ft. and 35,000 lb. weight	286 kt. TAS
Maximum range with 50 passengers, instrument flight fuel reserves at maximum cruise thrust	1,450 naut. mi.
Maximum range with 50 passengers, instrument flight fuel reserves at long-range cruise thrust	1,590 naut. mi.

ATR42 offers commuter airlines simple, efficient operations

David M. North/Toulouse
July 15, 1985

Aerospatiale/Aeritalia ATR42 offers regional airlines a 46-passenger aircraft that is simple and efficient to operate and has the ability to carry standard cargo containers. It has the largest seating capacity of the new generation of turboprop aircraft entering commuter operations.

This *Aviation Week & Space Technology* pilot flew three nights in two different ATR42s, built by the French and Italian aircraft manufacturers, during the evaluation from the Aerospatiale facility here. The No. 1 aircraft was flown to assess the handling qualities of the Pratt & Whitney Canada PWI20-powered commuter aircraft. The No. 3 flight test aircraft was flown on the first and third flights to compare the aircraft in commuter-type operations with previously flown aircraft.

The third ATR42 has been configured for airline operations and has been used to test the production aircraft cockpit displays, electrical system and the air conditioning. The No. 3 aircraft still has some minor aerodynamic changes to be incorporated prior to delivery to a

Aerospatiale has accumulated more than 800 hr. of flight testing in three Aerospatiale/Aeritalia ATR42 commuter aircraft since first flight on Aug. 16, 1984. The third of the 42-50-seat aircraft joined the flight test program on Apr. 30. Certification of the $7-million aircraft, powered by Pratt & Whitney Canada PW120 turboprop engines, is scheduled for mid-September.

customer. The No. 2 aircraft was in use by flight program test pilots at the time.

The ATR42 was flown with Gilbert Defer, chief test pilot at Aerospatiale. Prior to the first flight in the No. 3 aircraft, Defer detailed some of the features of the high-wing aircraft. One feature designed especially for commuter operations and for fast turnarounds is the use of the right engine as an auxiliary power unit. A propeller brake mounted on the reduction gear box allows the engine to be operated without propeller rotation. The brake is hydraulically actuated but does not require hydraulic pressure once engaged. The right engine at ground idle provides electrical and hydraulic power and also operates the air conditioning system for the cockpit and cabin. The fuel flow in the auxiliary power setting on the ramp was 176 lb./hr.

Aerospatiale and Aeritalia also designed the ATR42's systems to be accessible on the ground for routine maintenance. For example, the hydraulic system components are installed in the landing gear aft fairing within an average person's reach. The refueling panel is located in the right landing gear fairing.

The design simplicity sought by the two manufacturers was evident in the cockpit while I was sitting in the right seat observing Defer in the left seat conduct the pre-start check list. I sat in the right seat for the first flight to become acquainted with the instrument panel and system operation and occupied the left seat for the remaining two flights.

Defer completed the checklist in less than 1 min. All of the checklists for the ATR42 were the shortest I can remember for an aircraft of its size. Aerospatiale borrowed much of the system display technology from the Airbus transport aircraft. Defer had only to make certain that all lights were out on the overhead panel to assure that all systems were operating normally.

White lights denote an unusual position of a switch, blue lights indicate a temporary use of a system, red lights issue the usual warning signal and the amber shows caution. The pre-start checklist took slightly longer than normal to accomplish because it was the first flight of the day.

The ramp weight of the No. 3 aircraft, with registration F-WEGC, was 29,000 lb., or 84% of its maximum takeoff gross weight. Total fuel on board at the ramp was 5,470 lb., or 55% of its maximum fuel capacity. The No. 3 aircraft is to be certificated as an ATR42-200 with a maximum takeoff weight of 34,725 lb. The ATR42-300 will have a 35,605-lb. maximum takeoff gross weight.

Both aircraft have the same configuration, and the only difference between them may be in the certification and in the flight manual. Aerospatiale is conducting tests to determine whether there will be any changes needed in the landing gear system for the higher gross weight. The ATR42-300 was developed for Finnair to carry more fuel on some of its longer routes.

Engine start procedure for the Pratt & Whitney 1,800-shp. PW120 turboprop engines is straightforward. It is accomplished by depressing the start button on the overhead panel and then moving the condition lever to the idle position once the high-pressure rotor speed has exceeded 10%. The starting temperature was 830C. It took Defer 6 min. from engine start to reach the takeoff position on Runway 33 at Toulouse.

The first takeoff by Defer was an excellent demonstration of the single-engine capability of the ATR42. Shortly after the aircraft achieved a positive rate of climb and I had raised the landing gear, the propeller brake light illuminated, indicating the propeller brake system could be activated on the right engine. Defer said that the temperature limits on the brake system were too narrow and that the system was still being fine-tuned.

First delivery of an ATR42 is scheduled for October to the French regional carrier Air Littoral. Aerospatiale also will deliver the commuter aircraft to Denmark's Cimber Air and Finnair this year. Production rate of the ATR42 is planned to reach two monthly in early 1986 and increase to three monthly by the end of 1986. The maximum production rate of four ATR42s a month is expected in 1987. Aerospatiale has the responsibility for the construction of the aircraft's wing sections, cockpit layout, engine installation, flight testing and final assembly at its Toulouse facility. Aeritalia builds the ATR42's fuselage, including the tail.

A temporary direct-reading temperature gauge on the propeller brake had been installed to define the parameters but was not yet operational. He also said that the chances were 100 to one that the fault light was in error but that a precautionary engine shutdown was the best course of action.

An immediate downwind leg was requested and granted by the airport tower as the right engine was shut down. Defer maintained 30 deg. of flaps on the downwind, lowered the gear at the 180 deg. position and selected final flaps once we were assured of making the field. He flew the approach 10 kt. faster than the reference speed of 95 kt. for the aircraft's gross weight, and stopped the aircraft with a rollout of less than 2,000 ft. by using only the reverse pitch from the left engine and no brakes. Defer backed the ATR42 with the single engine to leave the runway at the first turnoff.

Approximately 11 min. elapsed from engine start to engine shutdown on the Aerospatiale ramp. During that time, 150 lb. of fuel had been used. Aerospatiale mechanics determined that the illumination of the propeller brake light had been a faulty indication. The brake system was reset and the direct-reading temperature gauge was activated as a backup indication for our final flight later in the day.

During the flight in the No. 1 ATR42, with French registration F-WEGA, I took the left seat and Defer the right. The gross weight of the first prototype ATR42 was 33,000 lb., including 4,000 lb. of fuel and 8,000 lb. of flight test equipment. There were two other people in the aircraft during this flight to evaluate the ATR42's handling qualities.

I flew steep turns during the climb to 6,000 ft. prior to performing stalls. The ATR42 was responsive to both pitch and roll inputs during the turns at higher speeds. The main flight controls are actuated manually by cables and rods without any servo-controls. Defer said that achieving a balance between aileron forces required for both the slow- and high-speed regimes and providing adequate rudder control during an engine-out situation without boosted controls were the most difficult projects in the flight test program.

The ailerons and elevator are statically and dynamically balanced on the ATR42. Small sets of hydraulically operated spoilers were installed in the aircraft's upper wing to lower the aircraft's minimum control speeds and to aid in lateral control. The spoilers, however, are not required for dispatch and hardly degrade the aileron control. They are helpful at approach speeds with an engine out, Defer said. The spoilers start to deploy with increasing aileron travel.

Aerospatiale and Aeritalia experimented with several configurations for the unboosted rudder. An initial rudder design using an auto tab and a balance tab was changed to a spring tab with the horn section above the horizontal tail part of the spring tab. The rudder also was enlarged to keep the minimum control ground speed at 85 kt. while requiring less than 132 lb. of rudder force.

The wing of the ATR42 was designed to enable the aircraft to fly at relatively high angles of attack at lower speeds. The double-slotted flaps along more than 60% of the wing span also aid in maneuvering at low speeds. At higher speeds, the roll rate for the ATR42 is close to 60-deg./sec., while at lower speeds the roll rate is reduced and the stick forces are increased.

At 6,000 ft., I pulled the throttles to idle and began a slow deceleration to stall speed. The angle of attack indicator in the cockpit showed that the ATR42 was at least 2 deg. beyond the maximum coefficient of lift when the aircraft fell off on one wing at close to 90 kt. without a nose drop.

Mild buffet

Defer said the company had considered installing a stall strip on the wing to achieve recognized stall behavior. As it was, a very mild buffet was encountered prior to the theoretical stall speed and then the wing dropped after the stall. Aileron control was available into the stab and for the recovery from the left-wing drop.

Defer said flight tests with a stick pusher were scheduled the following week. The pusher was not to be considered a limiting factor, Defer said, because it was to be programed to activate after reaching stall speed.

The aircraft exhibited the same stall characteristics but without the sharp wing drop with the landing gear down and the flaps in the landing configuration. The wing drop occurred at 70 kt., and again there was ample pitch control to nose the aircraft over and aileron control to attain wings-level attitude during the recovery.

Following the stall sequence, Defer retarded the power lever on the left engine. The amount of rudder force required initially to center the yaw indicator ball was calculated to be 120 lb. by the flight test engineer while flying at near 100 kt. with the landing gear extended and the flaps set at 30 deg. Full power on the right engine gave a 500-fpm climb at 6,000 ft. altitude. Balanced flight was maintained by the use of 5 deg. of bank and rudder trim. Control forces for turns, both into and away from the dead engine, were nominal.

High-speed descent

With the left engine restarted, both engines were placed at idle with the landing gear extended and the flaps set at 30 deg. A high-speed descent was established at 240 kt. that resulted in a rate of 4,000-5,000 fpm. The cockpit noise in the No. 1 ATR42 was higher than normal during the high-speed descent, primarily because the prototype aircraft was not equipped with all of the soundproofing to be installed on the production aircraft. Later flight in the No. 3 aircraft confirmed the quieter operation of the production-configured ATR42.

I flew the ATR42 in the landing pattern at the Toulouse airport at approximately 120 kt. in the clean configuration, slowing to 95 kt. for the approach. The aircraft was responsive to both pitch and roll control inputs during the approach and landing. A slight crosswind from the left was easily countered by a slight wing down and opposite rudder. Defer said the ATR42 would most likely be certificated for direct crosswind components of 40 kt. or greater.

The first landing was less than smooth as I searched for the correct attitude. The trailing link landing gear eased the touchdown somewhat. The aircraft was stopped in less than 3,000 ft. without the use of brakes and less than full reverse pitch. The power levers only have to be pulled straight back through a restraining stop to achieve reverse and do not have to be lifted up and back. The second landing was smoother.

Smooth touchdown

By the time I landed the ATR42 the third time, I was able to achieve a smoother touchdown and stop in less than 2,000 ft. by the judicious use of brakes and partial reverse power. The total fuel used during the 1.1-hr. flight from blocks to blocks was 1,180 lb. The total included 56 min. of air time, less 3 min. sitting on the runway during the stop-and-go landings. Two stop-and-go landings plus one final landing were made during the flight.

The weight of the No. 3 aircraft at the Aerospatiale ramp for the third flight was 29,000 lb., including 5,380 lb. of fuel. The aircraft's weight was 84% of its maximum takeoff gross weight and we were carrying 54% of the maximum fuel capacity. The third ATR42 was configured in a standard commuter configuration with 46 seats and the forward and aft cargo areas. For the third flight, a copilot and another passenger were on the aircraft with Defer and myself.

Engine start

Engine start on the right engine was performed prior to our boarding the aircraft for the last flight. The noise from the engine in the auxiliary power unit mode was not excessive while boarding the aircraft from the left side. Once I was in the left seat and Defer in the right, the left engine was started using normal procedures. Approximately 45 lb. of fuel was used from engine start to beginning of taxi. Another 25 lb. of fuel was consumed during the 3-min. taxi to the active runway.

Four Sperry electronic flight instrument system (EFIS) displays are standard equipment in the Aerospatiale/Aeritalia ATR42. The Sperry digital system, organized around an Avionics Standard Communication bus, includes a DFZ 600 automatic flight control system cleared to Category 1 minimums. As an option, the autopilot can be linked to the radio altimeter for Category 2 approaches. Aerospatiale offers a King Gold Crown 3 communication and navigation package as standard, but with Collins Pro Line 2 as an option. Aircraft system displays and controls are located in the overhead panel.

The hydromechanical nosewheel steering system on the ATR42, controlled from a handle on the left side console, seemed sensitive at first, but gradually I was able to make heading corrections with less overcontrol. The nosewheel steering was used for the initial portion of the takeoff roll until the rudder became effective. The rotation speed was 95 kt., and the landing gear was raised at 100 kt. after we had established an initial 15-deg. climb. A climb speed of 150 kt. was set after the takeoff flaps were retracted. The takeoff fuel flow was 880 lb./hr./engine and the takeoff roll was slightly more than 2,000 ft.

Passing through, 5,000 ft. from the airport—which has a 500-ft. elevation—the airspeed was 150 kt., and the fuel flow registered 730 lb./hr./engine. The initial rate of climb was close to 2,000 fpm and averaged 1,500 fpm over the 3 min. it took to reach 5,000 ft.

Another 3 min. were required to reach 10,000 ft., and the fuel flow had dropped off to 715 lb./hr./engine. The total fuel used from takeoff to 10,000 ft. was 160 lb., with another 70 lb. of fuel used during engine start and taxi. During climb, Defer engaged the Sperry DFZ 600 automatic flight control system and chose the climb speeds and the 16,000-ft. cruising altitude.

The Sperry flight control system leveled the ATR42 smoothly at 16,000 ft. Climb power was left on after attaining the cruising altitude to achieve maximum cruise airspeed. The total time from takeoff to 16,000 ft. was 9.5 min., and the fuel used was 258 lb. The fuel flow at the climb power was 617 lb./hr./engine while level at the cruising altitude.

Approximately 3 min. were required to attain a maximum cruise at 16,000 ft. of 205 kt. indicated. The torque at maximum cruise was 76%, and the engine temperature was approximately 720C. The true airspeed on the cooler-than-standard day was 264 kt. The fuel flow at the maximum cruise setting was 585 lb./hr./engine. Defer said he was normally able to achieve a 270-kt. maximum cruise speed in the 17,000 to 20,000-ft. range.

The power levers on the two Pratt & Whitney PWI20 engines were retarded to achieve a 240-kt. true airspeed, approximating a long-range cruise speed. The indicated airspeed at this power setting was 186 kt., and the fuel flow was 485 lb./hr./engine. The gross weight of the aircraft at cruise altitude was slightly more than 28,500 lb. The nominal cabin differential pressure for the aircraft is 6 psi. allowing a 6,700-ft. cabin altitude at 25,000 ft.

During the short cruise portion of the flight, I was able to observe the cockpit layout and the instrumentation of the ATR42. The experience gained from development of the Airbus Industrie A320 cockpit display by Aerospatiale was evident. The cockpit instrumentation was designed to minimize the workload for the two-person crew. The light-out concept, except for advisory messages, was augmented by clearly marked system diagrams on the overhead panel. The cockpit of the ATR42 has ample room to move around and access to the seats is more than adequate. The noise level in the production-configured aircraft at cruise speeds permitted discussion between myself and Defer at conversational levels.

EFIS integration

Integration of the four-tube Sperry electronic flight instrument system displays with the rest of the digital avionics systems in the ATR42 also indicate Aerospatiale's involvement in larger transport aircraft programs. Both copilot and pilot stations have Sperry attitude directional indicators and horizontal situation indicators. Weather radar displays can be superimposed on the horizontal situation indicator.

The Sperry EFIS displays were not standard in the early part of the ATR42's development program but recently have been incorporated into the standard avionics package. Category 1 approach minimums to 200 ft. are standard in the ATR42, but the autopilot system can be linked to the radio altimeter for Category 2 approaches, to a 100-ft. decision height as an option. The King Gold Crown 3 communication and navigation avionics package also is standard on the commuter aircraft, and the Collins Pro Line 2 system is an option.

Top of descent for the approach to Runway 13 at the Montpellier airport was reached 14 min. after achieving cruising altitude. The 115-naut.-mi. distance between Toulouse and Montpellier was chosen as a typical commuter leg for the ATR42. Defer was unable to talk with air traffic control because of the amount of radio traffic in the area, delaying the descent into Montpellier so that it had to be at a higher rate than normally used for commuter passenger operations. Defer said the high-speed descent compensated for the time lost taking long range speed and fuel calculations.

Initial descent speed was 220 kt. with a rate of 3,300 fpm. The rate of descent was increased to 4,400 fpm using a descent speed of 245 kt. indicated. The fuel flow during the descent was 175 lb./hr./engine. At 5,000 ft., I thought a 360-deg. turn to lose altitude might be required to avoid overshooting the runway, but Defer said we could land easily at the midpoint of the runway. Slowing the aircraft to an appropriate approach speed, I was able to make a relatively smooth landing in approximately the first half of the runway. The visibility from the cockpit during the descent and throughout the three flights was excellent. The visibility over the nose of the aircraft also is excellent, giving a good perspective of the runway during landing.

The landing took place 6 min. after departing the 16,000-ft. cruising altitude, and another 5 min. was required to reach the blocks at Montpellier, including a slow backup into a parking space on the airport's ramp. The block-to-block time for the 115-naut.-mi. leg was 37 min., aided by the high-speed descent, which would not be flown in normal flights for passenger comfort reasons, and a straight in approach to the active runway at Montpellier.

The air time for the leg was 29 min. and a total of 550 lb. of fuel was used during the airborne portion of the flight. The fuel usage for the 37-min., block-block flight was 658 lb.

Development of all five of the newer-technology commuter aircraft was begun in the aftermath of the deregulation of the U.S. airlines in 1978, when the forecast was for a rapidly expanding regional airline network. The passenger growth of the commuter airlines has been close to 15% annually in recent years, exceeding the expectations of many manufacturers. At the same time, there have been a number of bankruptcies, mergers and new entrants into the commuter carrier field within the past seven years, continuously changing the customer lists of the aircraft manufacturers.

Both the Saab-Scania/Fairchild Industries 340 and the de Havilland Aircraft of Canada DHC-8 entered commuter operations last year. Deliveries and orders of these two commuter aircraft through the summer of this year have not reached the numbers expected by the manufacturers at the time of aircraft roll out.

New aircraft flights

I flew the General Electric CT7 powered Saab/Fairchild 340 in May, 1984, and it was the first to enter service with Crossair in Switzerland and Comair in the U.S. (AW&ST June 11, 1984, p. 61). I flew the de Havilland DHC-8 in Canada earlier in the year (AW&ST Apr. 23, 1984, p. 137), and the Embracer EMB-120 Braslia in Brazil a month later (AW&ST May 28, 1984, p. 43).

The 36-passenger DHC-8 has been in commuter service since late last year and the Brasilia is scheduled to start flying in the U.S. with Atlantic Southeast Airlines late this month.

The ATR42 is the fourth of the new generation of commuter aircraft I have flown in the past 16 months. The one remaining new aircraft to be evaluated is the CASA/Nurtanio CN-235.

All four of the new-generation turboprop commuter aircraft I have flown have varying attributes. The initial price and the direct operating cost of the pressurized aircraft vary according to the number of passengers carried. The purchase price for the 30 passenger Embraer Brasilia is slightly less than $5 million. The initial cost of the 34 passenger Saab/Fairchild 340 and the 36 passenger DHC-8 both are in the $5.5 million category. The price of the standard-equipped 46 passenger ATR42 is $7 million. The aircraft manufacturers offer different financing packages.

The direct operating cost for each aircraft tends to parallel their size. The smaller Brasilia is generally less expensive to operate than the larger Saab/Fairchild 340, with the DHC-8 and the ATR42 still more expensive. The de Havilland aircraft pays a small penalty in fuel costs because of its configuration and nearly short-field takeoff and landing capabilities. Because of the 10-16 greater passenger capacity of the ATR42, compared with the three other commuter aircraft, it has a lower cost per seat mile for a 150-naut.-mi. trip, according to manufacturers' figures. The company estimates that the ATR42 has a break even load factor of 35% with 16 passengers on a 150-naut.-mi. trip.

Handling qualities and ease of performing commuter-type operations, both for maintenance and operationally, have been optimized for all four aircraft, with differences largely a preference of individual operators. Three of the evaluated aircraft are powered by Pratt & Whitney PW100 series turboprop engines, while the Saab/ Fairchild 340 has General Electric CT7 turbine engines. The CASA/Nurtanio CN-235 also is powered by a General Electric CT7 turboprop engine. The maximum cruise true airspeed of the ATR42, DHC-8 and the 340 are all in the 270-kt. range, with the smaller Brasilia having a 20-30 kt. advantage.

All four manufacturers of the new generation of commuter aircraft in the 30-46 passenger range offer electronic flight instrument systems (EFIS) as either standard equipment or an option. The Saab/ Fairchild 340 and the Embraer Brasilia come equipped with a Collins EFIS as standard equipment. Both the ATR42 and the DHC-8 carry the Sperry EFIS package, but the system is standard on the ATR42 and an option on the DHC-8.

Test hours

Aerospatiale has accumulated more than 850 hr. in the flight test program on three ATR42s since the No. 1 aircraft first flew on Aug. 16, 1984. The No. 3 commuter aircraft joined the test program on Apr. 30, 1985. French certification for the ATR42 is scheduled for mid-September and U.S. Federal Aviation Administration approval is expected to be granted by the end of the year.

The first delivery of an ATR42 is scheduled to Air Littoral in October or November, 1985. The Montpellier-based regional airline will receive the No. 3 ATR42 and has an order for another aircraft, both in a 50 passenger-seat configuration. The first scheduled delivery of an ATR42 to a U.S. regional airline is to Command Airways in January, 1986. Command, based at the Duchess County Airport in N.Y., has orders for five aircraft, two for delivery in 1986 and three in 1987, plus options for another two.

Aerospatiale expects to deliver at least three aircraft to customers this year, the first to Air Littoral and one each to Denmark's Cimber Air and Finnair. In early 1986, Aerospatiale will reach a production rate of two monthly, gradually increasing to three monthly by the end of the year. Production rate of the ATR42 is planned at four monthly in 1987, but there are no plans to go beyond that rate unless there is a large upsurge in orders.

The development cost through certification for the ATR42 closely resembles the costs expended by de Havilland on the DHC-8. Aerospatiale estimates that it will have spent close to $250 million in research and development on the ATR42 to reach certification. Embraer's development costs for the Brasilia were approximately $200 million.

The break-even point for Aerospatiale and Aeritalia is approximately 350 aircraft deliveries. Development costs for the Saab/Fairchild 340 were higher than the others, owing to the joint venture program, new facilities and establishing a marketing organization.

At this time, Aerospatiale holds firm orders for 54 ATR42s, and another 26 aircraft are covered by options. Aerospatiale's vice president of ATR42 marketing in North America, Alain Brodin, defines an order as an aircraft with more than a $100,000 deposit that is non-refundable. A total of 15% of the purchase price of the aircraft is due between contract signing and delivery. The aircraft manufacturer took options early in the program, but is no longer pursuing them, Brodin said.

Included among the airlines holding orders for the ATR42 are Simmons Airlines with four and Ransome Airlines with six. Wright Airlines, now in Chapter 11 bankruptcy, held orders for four. Air Midwest Skyways has money down on five aircraft, but the company is holding its contract with Aerospatiale dormant until the merger between Air Midwest and Scheduled Skyways is finished, when the company can assess its aircraft requirements. Finnair has orders for five ATR42s, Air Queensland for four, Alitalia's Aero Transporti Italiani for six and France's Brittany Air International for two.

The $7-million price for the ATR42 includes the standard 46-seat configuration with a 30-in. pitch, Sperry EFIS package, King avionics and the ability to fly to Category 1 minimums. Options include configurations with 40 passenger seats at a 32-in. pitch or 42 seats with a 31-in. pitch. Brodin said all of the U.S. airlines had chosen the standard 46-passenger-seat configuration. Another option offered by the company is a partition in the front of the cabin that can be either moved forward so that 50 seats can be accommodated or moved to the rear to create a front cargo area and still have 22 seats in the cabin.

A corporate version of the ATR42 also is available. The corporate version would differ primarily in the cabin interior, although the avionics options offered to operators would be expanded. Any ATR42 ordered by a corporate operator, other than the standard commuter configurations, would be delivered to a completion center for the installation of the interior.

Brodin said that most often in discussions with airlines his sales representatives are competing with the 340 and the DHC8, and not the 30-seat Brasilia. The Brasilia tended to complement the ATR42 in fleet planning for commuter operations, rather than compete against it, he said. All of the manufacturers of the new line of commuter aircraft in the 30-50-passenger category are competing for what is estimated by Aerospatiale and others to be a 2,800-3,000-aircraft market over the next 10 years. The initial level of orders for this class of aircraft is short of the expectations.

Aerospatiale has set up a separate company in the U.S. to handle product support for the ATR42. The facility, to be based at Washington's Dulles International Airport, is scheduled to open on Sept. 1. The spare parts inventory, technical representatives and warranty administration will be housed in the new facility. ATR42 pilot training will be conducted in Toulouse until 1987 when it will be moved to Washington for U.S. carriers. Mechanic training will be moved to the Dulles facility prior to 1987.

ATR42 SPECIFICATIONS

POWERPLANT

Two Pratt & Whitney Canada PW120 turboprop engines with a normal takeoff rating of 1,800 shp. An automatic power reserve feature enables each engine to be operated at a maximum of 2,000 shp. with one engine inoperative.

PROPELLERS

Two Hamilton Standard 14SF four-blade, 13-ft.dia. constant speed with full feather and reversible capabilities.

WEIGHTS

Maximum takeoff weight	35,605 lb. (16,150 kg.)
Maximum landing weight	35,275 lb. (16,000 kg.)
Maximum zero fuel weight	32,630 lb. (14,800 kg.)
Operational empty weight	21,986 lb. (9,973 kg.)
Maximum payload	10,644 lb. (4,827 kg.)
Maximum fuel load	9,920 lb. (4,500 kg.)

DIMENSIONS

Length	74.5 ft. (22.7 meters)
Height	24.9 ft. (7.6 meters)
Wingspan	80.6 ft. (24.6 meters)
Wing area	586 sq. ft. (54.5 sq. meters)
Cabin length from cockpit door to rear bulkhead	45.4 ft. (13.8 meters)
Cabin length of passenger compartment	31.4 ft. (9.6 meters)
Cabin height	6.3 ft. (1.9 meters)
Cabin width	8.5 ft. (2.6 meters)

PERFORMANCE

Balanced takeoff field length at maximum weight and standard conditions	3,740 ft. (1,140 meters)
Landing field length at maximum weight and standard conditions	3,215 ft. (980 meters)
Single-engine operating ceiling at 97% of maximum takeoff weight and ISA + 10C	8,750 ft. (2,667 meters)
Maximum range with 46 passengers, instrument flight fuel reserves at maximum cruise thrust	900 naut. mi.
Maximum range with 10 passengers, instrument flight fuel reserves at maximum cruise thrust	2,300 naut. mi.
Maximum cruise true airspeed at 20,000 ft.	270 kt.

Weight and performance figures are for the ATR42-300. The ATR42-200 is primarily the same aircraft as the ATR42-300, but with an allowable maximum takeoff gross weight of 34,725 lb. and a lower maximum zero fuel weight.

ATR42 SUPPLIERS AND VENDORS

Engines	Pratt & Whitney Canada
Propellers	Hamilton Standard
Air conditioning	Garrett
Pressurization	ABG Semca/Softair
Oxygen	Eros/Puritan
Fuel pumps	Bronzavia
Fuel valves	Le Bozec and Gautier
Fuel quantity	Intertechnique
Hydraulic pumps	Abex
Deicing boots	Kleber-Colombes
Ac/dc systems	Auxilec
Landing gear	Messier-Hispano
Brakes and wheels	Goodyear
Antiskid	Hydro-Aire
Flap actuators	Ratier
Cockpit windows	PPG
Crew seats	Ipeco
Passenger seats	Socea
Electronic flight instrument systems	Sperry
Automatic flight control system	Sperry
Avionics, nav and comm	King Gold Crown 3

SF-340 offers payload-range versatility

David M. North/Washington
June 11, 1984

Saab-Scania/Fairchild Industries 340, the first of the new generation of 30-50-passenger commuter aircraft to be certified, offers excellent handling qualities and various range and payload tradeoffs for regional airline and corporate operations.

The 34-passenger turboprop aircraft is a joint venture between the Swedish and U.S. aircraft manufacturers. It is the third of the new commuter aircraft to be flown by this *Aviation Week & Space Technology* pilot in the last two months. The de Havilland Aircraft of Canada DHC-8 was evaluated in early April (AW&ST Apr. 23, 1984, p. 137), and the 30-passenger Embraer EMB-120 Brasilia was flown in Brazil in late April (AW&ST May 28, 1984, p. 43).

Saab-Fairchild 340 No. 6, with Swedish registration SE-E06, was the aircraft used for the evaluation during flights from Washington's Dulles International Airport. The aircraft is scheduled to be delivered to Comair in mid-June and is painted in the colors of the Cincinnati-based regional airline.

Initial commuter operations of the Saab-Fairchild 340 by Crossair are scheduled to begin in Switzerland this month. U.S.-based Comair is scheduled to receive this aircraft—No. 6—in June, followed by the No. 4 aircraft in July. Comair has orders for 12 aircraft. Saab-Fairchild holds 60 firm orders for the corporate and commuter aircraft, and another 40 positions are covered by letters of intent and options.

I flew three flights in the commuter aircraft, which is powered by General Electric CT7-5 engines, during the same day. The first was flown in the right seat from Dulles to the Lynchburg, Va., airport, where the aircraft was demonstrated to an Air Virginia pilot. I flew the return trip to Dulles in the left seat and made a short flight to evaluate the handling qualities of the aircraft, also from the left seat. The first flight was flown by Fairchild Republic's director of flight operations, James Martinez, who assisted me from the right seat during the last two flights.

Prior to the first flight, Martinez detailed some of the servicing points for the aircraft and some of its exterior features. Maximum gross weight of the Saab-Fairchild 340 is 27,300 lb. The single-point refueling adapter and control panel is located in the leading edge of the right wing outboard of the engine nacelle. The ground air-conditioning and the external electrical connections are located in the right wing-to-fuselage fairing. Hydraulic servicing is through a connection in the nosewheel well.

The design of the Saab-Fairchild 340 has changed very little since the aircraft was given formal approval by both companies in January, 1980. The original concept for the aircraft showed a horizontal stabilizer in a cruciform configuration. Later in the year, the cruciform empennage was dropped in favor of a fuselage-mounted horizontal stabilizer with dihederal.

As with the Dash 8 and the Brasilia, the maximum gross weight and typical operating weight of the Saab-Fairchild 340 increased during its development phase. In early 1980, an optimistic maximum takeoff weight of 24,000 lb. and an operating empty weight of 14,600 lb. was forecast. Takeoff field length for the 24,000-lb. aircraft was estimated to be close to 3,000 ft.

In late 1980, the maximum gross weight of the Saab-Fairchild 340 was increased to 25,000 lb. and the operating empty weight to 14,700 lb. During construction of the first prototype in early 1982, the operating empty weight was increased to 15,860 lb. and the takeoff weight to 26,000 lb. Balanced field length was estimated to be 3,700 ft.

Saab-Fairchild increased the maximum takeoff weight to its present 27,000 lb. in 1983 after enlarging the lavatory and offering a 35-seat configuration as standard. The current typical operating empty weight is near 17,210 lb. and the balanced field length requirement is 4,000 ft. Payload for the aircraft has been maintained near 7,300 lb. during the development.

During their development, the larger Dash 8 grew 2,500 lb. and the smaller Brasilia 4,000 lb. from initial target values. The Saab-Fairchild 340, with a maximum takeoff weight between the Brasilia and the Dash 8, had a weight growth of 3,000 lb. As with the other two manufacturers, Saab-Fairchild increased the power of the engines, raising the rating of the General Electric CT7-5 engines from 1,630 shp. to 1,700 shp.

Early estimates of 340 type certification and deliveries in 1983 were made by the joint team in 1980. Since before the aircraft's rollout in October, 1982, and first flight in January, 1983, Saab-Fairchild officials have been predicting a March, 1984, certification; they missed the mark by only two months. The 340 received its Swedish type certificate on May 30 (AW&ST June 4, p. 29). U.S. Federal Aviation Administration certification is planned this month.

The layout of the cockpit and instrumentation has been designed for commuter operations. The uncluttered instrument panel has four 5-×-6-in. Collins Electronic Flight Instrument System (EFIS) tubes as standard equipment. Another optional multifunction display can be added for navigation, self-diagnostics and checklist presentation.

Aircraft system controls and indications are in the overhead panel, with circuit breaker panels located to the side of the pilots. The EFIS display control, altitude alert and navigation control panels are located in the eyebrow panel, with the communication controls within easy reach in the center pedestal.

The weight of the Saab-Fairchild 340 at the Page Airways ramp at Dulles was 23,000 lb., or 84.2% of its maximum ramp weight of 27,300 lb. Included in the total was 2,500 lb. of fuel, seven passengers and assorted bags in the aft cargo compartment.

Takeoff decision speed V_1 for the aircraft and the rotation speed was calculated by Martinez to be 106 kt. The V_2 takeoff safety speed was 107 kt. Martinez started the right engine first using the batteries in the fuselage fairing aft of the wing. The starting sequence is almost completely automatic and involves putting the condition lever in the start position prior to pushing the start button. The starter stops at 55% engine speed, and the engine's internal turbine temperature is then monitored to see that it does not exceed 920C. The temperature was 20C warmer than standard during the start, and the turbine temperature peaked at 743C.

During the taxi to Runway 30 at Dulles, the fuel flow was 200 lb./hr./engine, engine speed was 70% and the propeller speed 1,130 rpm. A total of 180 lb. of fuel was consumed during the 12 min. in the blocks and the 10-min. taxi. The aircraft does not have a total fuel burn indicator, so fuel measurements are not as accurate as obtained during the Dash 8 and Brasilia flights. Fuel flow on the 1,700-shp. CT7-5 engines during takeoff was 740 lb./hr./engine.

The balanced field length requirement for the Saab-Fairchild 340 at the warmer than standard temperature and with little wind was

3,100 ft. Actual takeoff roll was about 1,500 ft. The balanced field requirement for the Saab-Fairchild 340 is greater than that of the Dash 8, but less than the lighter-weight Brasilia.

A total of 3 min. was required to reach 5,000 ft. where the speed was 170 kt. and the fuel flow was 630 lb./hr./engine. Rate of climb passing through 5,000 ft. was 1,800 fpm. In another 2 min., Martinez was climbing the aircraft through 8,000 ft., the fuel flow had dropped to 600 lb./hr./engine and the speed was 166 kt. The rate of climb at the 8,000 ft. mark was 2,000 fpm.

Assigned altitude

Prior to reaching the assigned cruising altitude of 12,000 ft., the fuel flow had dropped to 520 lb./hr./engine. A total of 8 min. was required to reach 12,000 ft., and the total fuel used from takeoff was approximately 200 lb.

A maximum cruise torque setting of 70% with propeller speed of 1,300 rpm was selected at 12,000 ft. to give a maximum cruise speed of 210 kt. indicated. The ambient temperature was 5C warmer than standard, and the true airspeed was 255 kt. The fuel flow was 500 lb./hr./engine. Martinez said that during the 4,130 naut.-mi. flight from Sweden to Dulles, the average true airspeed was 265 kt. at 22,000 ft. using higher than long-range cruise speeds. Average cruise fuel flows during the 17-hr. 42-min. flight was 415 lb./hr./engine.

Descending through 9,000 ft. for the approach into the Lynchburg airport, the speed was 210 kt. and the rate of descent 1,500 fpm. The fuel flow during the descent was 330 lb./hr./engine at the 30% torque setting. Martinez landed the aircraft on Runway 03 at Lynchburg after a 36-min. flight. Another 2 min. was required to reach the Air Virginia ramp where the aircraft was to be displayed and flown by one of the airline's pilots.

Total fuel

A total of 800 lb. of fuel was used during the 1-hr. blocks-to-blocks operation, while 620 lb. was used during the actual flight. Subtracting 80 lb. of fuel used during the longer than normal time spent on the ramp at Dulles, the fuel burn for the flight for a typical commuter operation would be 720 lb.

The 120-naut.-mi. flight from Dulles to Lynchburg was approximately the same as the flight I had made in the Brasilia between two points in Brazil. The Saab-Fairchild 340 was flown 4,000 ft. below its optimum altitude of 16,000 ft., while the Brasilia was flown at its optimum altitude of 20,000 ft. The flight in the Saab-Fairchild 340 took

3 min. longer and burned 70 lb. more fuel, but had the Comair aircraft been flown at 16,000 ft., the fuel figures would have been closer.

The specific range for the Saab-Fairchild 340 at 20,000 ft. and 2,000 lb. less than maximum weight is 0.3 naut. mi./lb. of fuel. The speed at the maximum-speed cruise setting is 269 kt. For the Brasilia, under the same weight and altitude conditions, the specific range is 0.27 naut. mi./lb. of fuel, but the speed is 296 kt.

The fastest of the three new commuter aircraft is the Brasilia. The 340 is second in maximum cruise speed. The fuel burn on a given trip is very close between the 340 and the Brazilian-built aircraft, when both are flown near their optimum altitudes. The Dash 8 pays a penalty for its high wing configuration and near short takeoff and

Basic seating configuration for the Saab-Fairchild 340 is for two aft-facing seats in the front of the cabin and 11 rows of three-abreast seating, for a total of 35 passengers. Seating is at a 30-in. pitch; aisle width is 16.3 in. The seating arrangement selected by U.S.-based Comair has a galley and a luggage area for hanging clothes, replacing the two aft-facing seats.

landing performance in increased weight, resulting in a slower cruise speed and higher fuel flows than the other two aircraft.

Following a 45-min. flight by the Air Virginia pilot, I took the left seat and Martinez the right seat for the return to Dulles.

Automatic start

The left engine was started by the automatic procedure with electrical power from the operating right engine. The fuel at the blocks at the Air Virginia ramp was 1,120-lb. I made the takeoff from Lynchburg after a 3-min. taxi. The takeoff run was approximately 1,200 ft. for the 21,600-lb. aircraft, and rotation was positive to a 10-deg. nose-up attitude. The initial rate of climb was close to 3,000 fpm, dropping to 2,000 fpm passing through 10,000 ft. A total of 5 min. was required to reach the 10,000-ft. mark, and the speed was 145 kt.

As the aircraft passed through 13,000 ft., fuel flow was 520 lb./hr./engine, and the rate of climb had eased to 1,600 fpm. Total time from takeoff from Lynchburg to the 15,000-ft. cruising altitude was less than 9 min.

A maximum cruise power setting and 1,330-rpm propeller setting at 15,000 ft. yielded a true airspeed of 265 kt. and a fuel flow of 510 lb./hr./engine. Using the same propeller speed, the fuel flow was adjusted to 350 lb./hr./engine to yield a 215-kt. true airspeed at long-range cruise.

During the descent into Dulles, I increased the speed to 250 kt. and the rate of descent to 3,000 fpm. The noise level in the cockpit was low enough for Martinez and me to talk at normal conversation levels. The Saab-Fairchild 340 was equipped with soundproofing and interior furnishings while both the Brasilia and Dash 8 aircraft I flew were prototypes without any insulation.

Lateral control of the 340 at the higher speeds during the descent was excellent and approximated that of the Brasilia. Pitch control for the aircraft was positive at the higher and lower speeds, enabling me to place the nose at any desired point without straying. Both the 340 and the Brasilia appeared to have the same performance in pitch and roll control. The Dash 8 has a faster roll rate at slower speeds due to wing spoilers.

Visibility from the left seat, as from the right seat, is excellent, enabling me to look down the nose of the aircraft. Without undue straining from the pilot's seat, I was able to look back to the left engine nacelle and propeller spinner.

A long straight-in approach to Runway 1R at Dulles was flown at 115 kt. once the landing gear was extended and flaps were set at the 35-deg. landing position. I found the aircraft to be stable in the approach configuration, and once power was set for a constant descent, few power changes or control inputs were required. The end of the runway was crossed at 105 kt., and the aircraft touched down at approximately 90 kt.

The total time for the 120-naut.-mi. return trip to Dulles was 50 min., of which 42 min. was actual flight time. It took longer for the return trip because some slower speeds were flown during the cruise segment and a longer approach pattern was used into Dulles. Fuel burn for the return flight was 670 lb.

While parking the aircraft at the Page Airways ramp, I found the nosewheel steering to be effective in making tight turns to maneuver around other aircraft on the crowded ramp. The nosewheel turns 60 deg. from the center and is controlled by a wheel mounted to the left of the pilot's seat. Nosewheel steering also was used during the initial takeoff roll, until the rudder becomes effective above 40 kt.

The third flight was made in the early afternoon with 3,200 lb. of fuel in the aircraft. Ramp weight was approximately the same, 23,000 lb. Three passengers were carried, fewer than on the earlier flight, and their weight was offset by increased fuel.

I had performed the takeoff at Lynchburg from a full stop, adding takeoff power prior to releasing the brakes. The takeoff from Dulles was made while rolling from a high-speed taxiway. During both takeoffs, I advanced the throttles to near takeoff power and let Martinez set the precise power. Both Martinez and I monitored the torque setting and found only a slight increase in the setting during the takeoff roll.

Saab-Fairchild is installing a modified electronic control unit to limit torque and turbine temperature in its aircraft. The Comair aircraft was not equipped with the new unit, but I found that setting power was similar to other turboprop aircraft I have flown. The torque limiter was required during the certification of the 340 to correct for earlier torque rises during takeoff. The new control unit also will allow for full reversing power to be available during landing.

I flew the 340 to 12,000 ft. to the west of Dulles to perform slow flight and stalls. During the transition from faster to slower speeds and back to faster, I found that I was continually changing rudder trim to keep the yaw ball centered. Martinez said the company plans to install an automatic rudder trimmer within the next year.

Deployment of initial flaps prior to extending the landing gear produced a ballooning effect, requiring considerable forward pressure on the yoke to maintain altitude. The increase to full flaps did not have the same effect on performance. The same degree of attitude change with deployment of initial flaps was not noted in the Brasilia or Dash 8.

A stick shaker and a stick pusher are installed in the 340. The Brasilia has a stick pusher and shaker, while the Dash 8 has only a stick shaker. The stick pusher was required, Martinez said, because the aircraft has a tendency to roll to the left beyond limitations, when stalled with the flaps in the 20-deg. position.

Stall speed

In the clean configuration with power at idle, the stick shaker activated at 95 kt. and the stick pusher nosed the aircraft over at 88 kt. Martinez said stall speed under these conditions would be 83 kt. Full stalls were not possible, Martinez said, because of the baggage and cargo in the rear hold that could be damaged.

The aircraft shaker started at 82 kt., with the landing gear extended and the flaps at 35 deg. At 12,000 ft. altitude and with an idle power setting, the stick pusher activated at 76 kt. Stall speed would have been 72-73 kt. if the aircraft were allowed to carry through to the stall. Using full climb power with the landing gear extended and full flaps, the stick shaker started at 70 kt.

The aircraft was nosed over by the pusher at 63 kt. while indicating a 1,300-fpm climb. Aileron control was available throughout the maneuver, and there was no tendency of the wing to drop at the stick pusher speeds.

Pitch and roll control at the slower speeds were positive and immediate. At the slower speeds, and with the landing gear and flaps extended with high power settings, the yaw experienced in the 340 was no greater than that in the Brasilia. More than adequate rudder was available to counteract yaw at full power.

Martinez shut down the left engine at 12,000 ft. A true airspeed of 190 kt. was maintained by the application of 85% maximum torque on the right engine. A total of 4.25 units of rudder was required to zero the yaw of the aircraft at cruise speed, and more rudder was available.

The left engine was restarted, and we began a descent into the landing pattern at Dulles. I performed one touch-and-go landing and a final full-stop landing on Runway 1R. Both approaches were visual with a close-in pattern to the runway requested by Dulles tower control. The turn from the 180-deg. position was tight with a short straightaway. I was comfortable with the handling qualities of the aircraft during both approaches, and both landings were relatively smooth. During the final landing rollout, I used partial reverse pitch on the propellers, and found that selecting reverse was easier in the 340 than in some other turboprops I had flown. During the 27-min. flight—40 min. from blocks-to-blocks—600 lb. of fuel were consumed.

Comair expects to receive the No. 6 aircraft in mid-June, following Federal Aviation Administration validation of the Swedish type certificate for the aircraft. The No. 4 Saab-Fairchild 340, also a Comair aircraft, is being used for function and reliability testing and is scheduled for delivery in July to the airline. The first operational Saab-Fairchild 340 is scheduled to go into service with Crossair from its Swiss bases in June.

Crossair has firm contracts for 10 of the aircraft, and Swedair has contracts for a like number. Comair has committed for 12 340s and Wichita-based Air Midwest has contracts for five.

The Comair aircraft flown from Dulles was configured with a large galley and a hanging closet on the right side of the cabin and 11 rows of three-abreast seating for a total of 33 passenger seats. The galley and storage closet can be removed, and two aft-facing seats can be installed in their place for a 35-seat configuration. A 34-seat configuration with a single aft-facing seat also is possible. The lavatory is in the rear of the cabin.

Airline interiors are being completed for the aircraft at Metair in England. Executive configurations for the aircraft will be done at Fairchild Aircraft's San Antonio facility.

The first executive version of the twin turboprop is to be the No. 14 aircraft, and it is scheduled for delivery to the Mellon Bank in October. The corporate aircraft will have a 16-passenger configuration and a Garrett auxiliary power unit. Phillip Morris Corp. has orders for three 340s in the corporate configuration.

Firm contracts

Saab-Fairchild has firm contracts for approximately 60 of the 340s in the commuter and corporate configuration. Another 40 aircraft are held under options and letters of intent. The partnership intends to deliver 24 aircraft in 1984 and 50 each in 1985 and 1986. The Swedish production facility has the capacity to build the aircraft at a rate of nine monthly, but there are no plans to reach that number in the near term.

Included in the 24 aircraft to be delivered in 1984 are open slots for three aircraft for purchasers with an immediate need for an aircraft. There also are some open slots in 1985 for the same reason.

The January, 1984, price for the regional airline version of the aircraft is $5.48 million, which includes the four-tube Collins EFIS. The multifunction display is an option. The corporate 340 has a list price of $6.30 million, which includes a standard executive interior and the Garrett auxiliary power unit.

The Saab-Fairchild Finance Corp. has been able to offer financing for the 340 at near 10% interest and for 10-12 years, Philip G. Dunnington, managing director of Saab-Fairchild, said. The company requires a $50,000 deposit and 5% deposits at six-month intervals so that at the time of delivery the customer has paid 25-30% of the cost of the aircraft. The internal finance company, backed by Citibank, has been given $125 million initially to finance purchases of the aircraft.

SAAB-FAIRCHILD 340 SPECIFICATIONS

POWERPLANT

Two General Electric CT7-5A turboprop engines with a takeoff and maximum continuous rating of 1,700 shp. An automatic power reserve feature is not included in the engine.

PROPELLERS

Two Dowty-Rotol R320/4-123-F/a four blade, 126 in. dia., constant speed with full feather and full reversible capabilities.

WEIGHTS

Maximum ramp weight	27,300 lb. (12,383 kg.)
Maximum takeoff weight	27,000 lb. (12,247 kg.)
Maximum landing weight	26,500 lb. (12,020 kg.)
Maximum zero fuel weight	24,500 lb. (11,113 kg.)
Operational empty weight	17,000 lb. (7,711 kg.)
Maximum payload	7,500 lb. (3,402 kg.)
Usable fuel	5,900 lb. (2,676 kg.)

DIMENSIONS

Length	64.8 ft. (19.7 meters)
Height	22.5 ft. (6.9 meters)
Wingspan	70.3 ft. (21.4 meters)
Wing area	450 sq. ft. (41.8 sq. meters)
Wing aspect ratio	11
Cabin length	34.7 ft. (10.6 meters)
Cabin width	7.1 ft. (2.2 meters)
Cabin height	6 ft. (1.8 meters)
Standard cabin configuration	34/35 passengers

PERFORMANCE

Takeoff field length requirement at 27,000 lb. with standard conditions and zero wind	4,000 ft. (1,219 meters)
Landing field length requirement at 26,500 lb. with standard conditions and zero wind, Part 25 requirements	3,610 ft. (1,100 meters)
Maximum cruise true airspeed at 25,000 lb. and at 20,000 ft.	269 kt.

Maximum range with 35 passengers with a fuel reserve for 100 naut. mi. diversion and 45 min. holding	800 naut. mi.
Range with maximum usable fuel and corresponding payload to reach 27,000 lb. with same fuel reserves	1,700 naut. mi.

Note: Average passenger weight for range calculations is based on 190 lb. each. Using a 200-lb. weight for each passenger, the maximum range would be 700 naut. mi.

Brasilia readied for commuter market

David M. North/Sao Jose dos Campos
May 28, 1984

Embraer is offering regional airlines a commuter aircraft with a standard 30-passenger configuration, excellent handling characteristics, high cruise speeds and attractive financing in its EMB-120 Brasilia, scheduled for initial deliveries beginning in mid-1985.

The Brazilian-built aircraft is the second of the new generation of 30-50-passenger commuters to be flown by this *Aviation Week & Space Technology* pilot. Earlier, I flew the de Havilland Aircraft of Canada DHC-8 (Dash 8), evaluated last month (AW&ST Apr. 23, 1984, p. 137).

FAA certification

Both the Brasilia and the Dash 8 are scheduled to be certificated to Federal Aviation Administration regulations Part 25 through Amendment 54. Both aircraft exhibit excellent handling qualities for their commuter role. The aircraft differences are evident in lateral control at different speeds, airport performance, fuel usage and cruise speeds. Both the 30-passenger Brasilia and the 36-passenger Dash 8 are powered by Pratt & Whitney of Canada PW100-series turboprop engines.

Embraer EMB-120 Brasilia is powered by two Pratt & Whitney of Canada PW115 turboprop engines rated at 1,590 shp. each. The Brazilian aircraft manufacturer has installed two strakes on the fuselage to improve the aircraft's single-engine climb performance. First of the $4.6-4.7-million 30-passenger commuter aircraft is scheduled to be delivered to Provincetown-Boston Airline in May or June, 1985.

The No. 1 prototype Brasilia, with registration PT-ZBA, was flown from Embraer's main production facility. The flight was with Flavio de Carvalho Passos, a major in the Brazilian air force assigned to Embraer for the EMB-120's flight test program in the right seat, while I occupied the left seat. Mauro Cesar Mezzacrappa, an Embraer flight test engineer, flew in the jump seat to monitor the aircraft's performance. The flight was planned in a similar manner to the commuter-type flight pattern that was flown in the de Havilland Dash 8. The first leg was flown to the Santa Cruz air force base, calculated to be 120 naut. mi. by way of the traffic route between here and the air force base near Rio de Janeiro. The actual route flown was approximately the same as that planned.

During the preflight inspection of the No. 1 prototype, Passos detailed a few of the aircraft's external features. The aircraft's systems are easily accessible during servicing on the ground, similar to Embraer's military aircraft. The avionics compartment, battery access, electrical power receptacle and hydraulic panel are in the nose. The single-point refueling system is under the right wing and outboard of the engine nacelle.

One of the few changes in the Brasilia since its first flight in July, 1983, has been redesign of the exhaust duct for the two Pratt & Whitney PW115 engines. The exhaust duct now extends back to the maximum chord of the wing and a tighter seal has been incorporated to improve aircraft performance. The redesigned exhaust duct increased the aircraft's rate of climb by 300 fpm and increased aircraft speed so that its maximum cruising true airspeed is close to 300 kt., Guido F. Pessotti, Embraer technical director, said.

Access to the pilot seats is easy, since the seats slide sideward as well as forward and rearward. Once I was in the left seat with adjustments made for my eye height and rudder pedal positioned comfortably, the aircraft's systems and controls were reached easily.

The aircraft's system controls and indicators are on the overhead panel with the circuit breaker panel to the rear of the system panel. The No. 1 prototype was fitted with a Collins electronic flight instrument system 4-x-4-in. attitude and horizontal situation indicators, standard on the Brasilia. A larger Collins display is optional, and Embraer intends to offer Bendix and King electronic displays as options.

Weather radar

Below the engine instrument and radio control panels on the center instrument panel and in front of the power lever quadrant is the multifunction weather radar tube and fuel flow and quantity indicators. The pressurization system controls and space for an area navigation system are in the control stand to the rear of the power lever quadrant. The cockpit is well designed with controls and system indications used more often during flight in position for easy access or observation.

Brasilia gross weight at the Embraer ramp was calculated by Mezzacappa to be 22,800 lb., and the center of gravity was 27.8%. Ramp weight included 4,000 lb. of fuel, 660 lb. of water ballast and an allowance for 600 lb. for two pilots and the test engineer. Aircraft ramp weight was 95% of the maximum 23,986 lb. that will be certificated for the Brasilia.

As in many aircraft programs, the basic operating weight and the maximum ramp weight of the Brasilia have increased beyond the target weights. An original ramp weight of close to 20,000 lb. and an operating weight of 11,620 lb. had been forecast by Embraer for the Brasilia in 1980. These weights were increased to design goals of a ramp weight of 21,340 lb. and a basic operating weight of 12,293 lb. in 1982 after Embraer had an opportunity to better evaluate the weight of the aircraft and its components.

Operating weight

The 2,775-lb. increase in basic operating weight from the design value has been mostly offset by the 2,646-lb. increase in ramp weight with little loss of payload and performance figures that nearly match Embraer's forecast for the aircraft at the lower gross weights. Embraer compensated for the increased gross weight by boosting the power of the PW115 engines from a planned 1,500 shp. to 1,590 shp.

Both engines were started by Passos using the engines' starter generator and external power. The engines could have been started by battery power or an auxiliary power unit, if installed. The No. 1 prototype did not have the optional Garrett GTPC36-150A tail-mounted auxiliary unit, although Embraer officials said most operators are installing the units for cooling, communication and for external lights at night.

The condition lever was moved to the fuel-on position once the high-pressure rotor speed had exceeded 10%. Starting temperatures on the 21C (69.8F) day were well below the maximum 810C (1,490F) interturbine temperature limit.

Another change being incorporated into production Brasilias is a back-up electrical power installation to smooth transition from external or auxiliary power to internal power. Embraer will install a Jet Electronics & Technology, Inc., power package to aid this.

I added power to taxi the Brasilia to Runway 15. During the taxi, Passos requested that I try the nose wheel steering prior to takeoff. The nose wheel has a 50-deg. deflection from centerline and is controlled by a steering handle on the pilot's left. Full control of the nose wheel is activated by depressing the pie-shaped handle. During 360-deg. turns on one of the ramps, I found the nose wheel steering positive without being overly sensitive. A trigger switch on the ram horn type yoke allows for 6-deg. of turn from center for takeoff and landing.

Fuel flow of the engines in the idle position was 150 lb./hr./engine with a high-pressure rotor speed of 53.2% and a propeller speed of 64%. Overspeed checks of the engines and Hamilton Standard propellers were accomplished while stopped, a procedure that Passos said normally would be done on the first flight of the day.

The V_1 takeoff decision speed was calculated to be 101 kt., the same as the V_r rotation speed. Takeoff safety speed, V_2, was 106 kt. on the morning flight with a 500-ft. overcast in rain.

A total of 210 lb. of fuel was used in the 21 min. from engine start to the beginning of takeoff roll. Because of the time spent in briefing and experimenting with the nose wheel steering, more fuel was used than would have been on a commuter-type operation from ramp to the runway. Passos said that in normal flights taxi fuel would have been approximately 80-100 lb.

The full takeoff torque of 106.7% was applied while holding the position on the runway with the brakes. Embraer does not have a torque limiter installed in the engines for takeoff but does allow for a 4% increase in torque during the takeoff roll.

After 1,400 ft. of takeoff roll, I rotated the nose to an 8-deg. attitude and the aircraft accelerated to 145 kt., once the landing gear and flaps were raised. The takeoff performance was affected by the increase in gross weight from the design weight. Embraer had expected a 3,500-ft. takeoff distance at the lower planned gross weight, and is

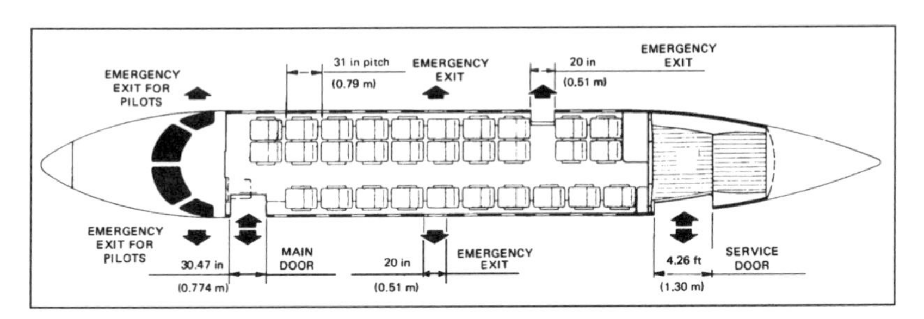

Standard interior configuration for the Embraer Brasilia is for 30 seats in three-abreast seating. The lavatory is located forward of the rear bulkhead on the aircraft's right side and across from the galley. The 226-cu.-ft. baggage area is reached through a 4.5-ft.-high and 4.3-ft.-wide cargo door. The rear bulkhead can be moved forward to increase the baggage area for less seating capacity. Interiors of the commuter version for the Brasilia will be done outside Brazil.

now expecting a 4,300-4,500-ft. takeoff requirement at maximum takeoff gross weight. The takeoff field requirement is greater for the Brasilia than for the Dash 8 with its almost short field takeoff and landing characteristics.

Rate of climb

The rate of climb initially after takeoff was in excess of 2,000 fpm, taking less than 2 min. to reach 5,000 ft. from the 700 ft. elevation of the Sao Jose dos Campos Airport. Passing through 5,000 ft., the fuel flow was 790 lb./hr./engine and the airspeed was 145 kt.

Another 4 min. was required to reach 12,000 ft., with the rate of climb at 1,800 fpm and the fuel flow indicating 660 lb./hr./engine. The rate of climb was once again at 2,000 fpm passing through 16,000 ft. and the fuel flow had dropped to 625 lb./hr./engine.

The assigned cruising altitude of 20,000 ft. was reached 11 min. after takeoff and with a total of 230 lb. of fuel consumed from take-off. During the later stages of the climb, slight to moderate turbulence was encountered, which imparted a noticeable yaw to the aircraft's tail. The No. 1 prototype is to have a new autopilot but did not have one during the flight. A yaw damper function of the autopilot would have reduced the tail yaw. Generally during the turbulence, the air-craft felt heavier than its 22,300-lb. gross weight.

During the climb into the overcast and through to the top level of the clouds at 16,000 ft., I was forced to adjust my cockpit scan for instrument conditions. The Collins electronic flight instrument system was flown easily with only minor variations in attitude and heading.

A maximum torque setting of 77% and an 85% rpm propeller set-ting at 20,000 ft. yielded a 198-kt. indicated airspeed. The true airspeed was calculated by Mezzacappa to be 298 kt. The temperature was al-most 20C (68F) warmer than a standard day. Fuel flow at maximum cruise thrust was 495 lb./hr./engine. The 10,000-lb. lighter weight Brasilia has a 2030 kt. faster cruise speed than the de Havilland Dash 8 and a lower fuel flow at comparable altitudes.

The power levers were then retarded until 56% torque was achieved and the propeller speed indication remained at 85%. Fuel flow at long-range cruise speed was 400 lb./hr./engine and the true airspeed was calculated to be 260 kt.

Embraer estimates that the Brasilia can fly three 100-naut.-mi. legs with 30 passengers without refueling if the aircraft is equipped with an auxiliary power unit. The fuel reserves for the three legs allow for a 100-naut.-mi. diversion with an additional 45 min. of fuel.

Without an auxiliary power unit, the Brasilia can fly four 100-naut.-mi. legs with the same fuel reserves. In both instances, the ground maneuver time at each airport is 8 min. using 77 lb. of fuel.

Cockpit noise

During the descent to the Santa Cruz military airport, I removed the radio headset and found that cockpit noise level was high at the 220-kt. descent speed. The No. 1 prototype does not have any soundproofing or insulation material. An air leak around the front passenger door contributed to the noise level. Passos said that the door and latch have been redesigned so that production aircraft will not have this difficulty and that the noise level in the No. 2 prototype is much lower than in the first aircraft. Descending through 10,000 ft. at 220 kt., the rate was 3,500 fpm at a torque setting of 10%. Fuel flow was 250 lb./hr./engine at the 10,000-ft. level.

Collins electronic flight instrument system and weather radar are standard on the Embraer Brasilia. The instruments located above the glare shield, a control box on the center panel and the circuit breakers to the right of the copilot are test instruments and will not be installed on production aircraft. The nose wheel steering handle is to the left of the pilot's seat and is controlled easily by resting the left arm on the armrest and depressing the handle. The 30-passenger EMB-120 commuter aircraft has a Grimes aural warning unit installed that uses a woman's voice to annunciate the warning that also is given by illuminated lights.

The visibility from the left seat was excellent, as a constant watch was kept for Northrop F-5Es of the Brazilian air force operating in the area during the approach. The wraparound windshield in the EMB120 affords an excellent view and a good perspective of the aircraft in relation to the ground during landing.

A speed of 110 kt. was flown in the landing configuration with a reference speed of 103 kt. at a weight of 22,200 lb. There is little pitch change with the deployment of the landing gear and initial flaps.

The electric elevator trim on the yoke was used to compensate for the selection of full flaps for the landing. The landing at Santa Cruz was firm, but not hard, as I searched for the correct eye reference to landing gear height.

The elapsed time from takeoff to landing at Santa Cruz was 33 min., during which the Brasilia was flown at both maximum and long-range cruise speeds. Total fuel used from takeoff to landing was 510 lb., and from start of engines to landing was 720 lb.

Subtracting 110 lb. of fuel used in evaluating the performance of the nose wheel steering at Sao Jose and adding 40 lb. for taxi at Santa Cruz, total fuel used for the 120-naut.-mi. leg was 650 lb. As with the Dash 8, the PW115s on the No. 1 Brasilia are preproduction engines, and the fuel flow and speed performance can be expected to improve slightly with production engines.

During the return flight to Sao Jose dos Campos, I climbed the EMB-120 to a clear area at 15,000 ft. to conduct hard turns and stalls. The Brasilia was very agile in pitch control and appeared to be more responsive in roll control at higher speeds than the Dash 8. The Dash 8, however, appeared to be more responsive in roll control at the slower speeds, with the aid of spoilers not on the Brasilia.

A maximum cruise thrust setting of 86% torque at 15,000 ft. resulted in an indicated airspeed of 216 kt. and a fuel flow of 570 lb./hr./engine. The true airspeed at the 15,000-ft. altitude was calculated to be 285 kt. At 216 kt., the aircraft was trimmed for hands-off flight and the aircraft remained stable for more than 4 min. in calm air without control input.

I reduced the power to idle at 15,000 ft., the stick shaker activated at 110 kt. and the aircraft stalled in the clean configuration at 102 kt. The stick pusher lowered the nose at 102 kt., approximately at the same time the aircraft stalled. The Brasilia has a natural stall warning with a slight buffet prior to stall. Embraer's Pessotti said the company was not willing to sacrifice aircraft performance to have the stick pusher removed.

With the flaps extended 15-deg., the stick shaker started at 93 kt. and the nose was lowered at 84 kt. by the pusher. Recovery from the stalls was accomplished by adding partial power, and the aircraft had no tendency to fall off on a wing during the stall. Aileron control was available throughout the stall and recovery.

The aircraft's stick pusher activated at 79 kt. following initiation of the stick shaker at 86 kt. with the flaps in the 25-deg. position and the landing gear extended. With the power still at idle, the stick shaker started at 82 kt. and the stick pusher moved at 78 kt. with the flaps in the 45-deg. position. The setting of 40% torque resulted in a stick shaker activation at 79 kt. and a stall at 74 kt., with a yoke that was almost to its full rear travel limit. As in the earlier stalls, the power-on stall with 45-deg. of flaps and extended landing gear was recovered easily with no unusual movements of the aircraft.

Passos retarded the throttle of the right engine and feathered the propeller to simulate an engine-out condition. At the 14,000-ft. altitude, the 100.3% torque setting of the left engine enabled the aircraft to maintain a 183-kt. cruise speed in level flight. The Brasilia required less than one unit of rudder trim to keep the yaw ball in the center. Service ceiling for the Brasilia on one engine is close to 17,000 ft.

During the stall and engine-out sequence, the pressurization system had no tendency to surge with the application or lowering of power. The Brasilia has a 7-psi. pressurization system that allows for a sea-level cabin up to 17,000 ft. and an 8,000-ft. cabin altitude at 32,000 ft. Embraer expects the service ceiling of the Brasilia to be 28,200 ft. with both engines operating. The Brasilia is to be certificated for a maximum operating altitude of 25,000 ft. to maintain the requirement for a manual release of oxygen masks. The executive version of the aircraft will be cleared to 32,000 ft., requiring an automatic release feature for oxygen masks.

Embraer recently incorporated several aerodynamic changes to the Brasilia, all of which were in the No. 1 prototype. Two strakes were installed on the lower side of the rear fuselage to decrease the side slip angle of the aircraft at maximum rudder deflections and to improve single engine climb performance.

Vortex generators and a small wing fence were added to the top of the wing to improve effectiveness of the ailerons during flight with the flaps extended. Another set of vortex generators was installed on the vertical tail below the horizontal tail to alleviate a pitch down moment during flight at a high sideslip.

Full-stop landing

A touch-and-go landing and a full-stop landing were made at the Sao Jose dos Campos Airport with smoother results than the earlier landing at Santa Cruz. The Brasilia was responsive to control inputs in the landing pattern and only minor power adjustments were required to hold a steady rate of descent for landing. On the final landing, the PW115 engines were put to almost full reverse and the aircraft was stopped in less than 1,400 ft. without use of full brakes. The landing distance for certification is 4,460 ft. at maximum landing weight.

During the 2 hr. 16 min. total time from block to block at the Embraer manufacturing facility, the fuel used was 1,960 lb. This included 1 hr. 40 min. of flight time and three landings.

The Brasilia has been designed so that a 96-in. plug could be installed in the fuselage and the seating increased by three more rows to a total of 39 passengers. The power required for a stretched Brasilia would be approximately 1,800 shp., Pessotti said, and the aircraft would have a cruise speed 20 kt. faster than the current Brasilia.

Embraer is planning to obtain Federal Aviation Administration certification for the Brasilia this year. The schedule for certification has slipped 4-5 months, primarily because the first prototype was delayed in construction.

The No. 1 prototype will support flutter testing as well as system testing. The initial prototype had 350 hr. at the time of the evaluation flight. The No. 3 prototype has accumulated more than 200 hr. and was used to complete icing and exterior noise testing in the U.S. in April. The No. 3 aircraft first flew in December, 1983. The third flying aircraft, the No. 4 Brasilia, flew early this month. The fourth prototype will be flown for systems development and then for hot-temperature and high-altitude testing. This aircraft is scheduled to be the first outfitted with a 30-passenger interior. The No. 2 Brasilia airframe will enter static testing in June followed by the No. 5 airframe to undergo fatigue testing here in September.

Production rate

First delivery of the Brasilia is scheduled for May or June, 1985, to Provincetown Boston Airline. The regional airline is expected to place firm orders for 10 EMB-120s and options on another five. Embraer plans to deliver 12 aircraft in 1985 and 24 in 1986. The manufacturer intends to maintain a two-month production rate of the Brasilia until the company can determine the market for the 30-passenger aircraft.

Embraer has the tooling and space to build the Brasilia at a monthly rate of six aircraft. At this time, the break-even point for the Brasilia is approximately 140 aircraft. A total of 118 options are held for the Brasilia by regional airlines and the Brazilian air force. Embraer is attempting to convert options to orders with the planned 16-month time lapse between orders and delivery. Other regional airlines in negotiations with Embraer for the 30-passenger aircraft include Air Virginia and a number of Brazilian carriers. Atlantic Southeast Airlines has orders for 10 and options for five Brasilias. The Brazilian air force holds options for 24 Brasilias for delivery as troop and cargo aircraft.

One of the attractive features of the Brasilia will be the financing package offered by Embraer. The manufacturer is quoting 10-year financing at 8% interest for purchasers of the EMB-120. The finance package includes a 15% deposit, but that can be relaxed to 10% depending upon the qualifications of the customer, Ozilio Carlos da Silva, commercial director of Embraer, said. A six-month grace period is extended for the first interest payment and a one-year grace period for the first payment of principle is being proposed by Embraer for Brasilia purchasers.

EMB-120 BRASILIA SPECIFICATIONS

POWERPLANT

Two Pratt & Whitney of Canada PW115 turboprop engines with a takeoff and maximum continuous rating of 1,590 shp. The maximum climb and cruise rating of the engine is 1,500 shp. An automatic power reserve feature is not included in the engine.

PROPELLERS

Two Hamilton Standard 14RF-9 four-blade, 10.5-ft.-dia., constant speed with full feather and full reversible capabilities.

WEIGHTS

Maximum ramp weight	23,986 lb. (10,880 kg.)
Maximum takeoff weight	23,810 lb. (10,800 kg.)
Maximum loading weight	23,260 lb. (10,550 kg.)
Maximum zero fuel weight	22,266 lb. (10,100 kg.)
Operational empty weight	15,068 lb. (6,835 kg.)
Maximum payload	7,198 lb. (3,265 kg.)
Usable fuel	5,755 lb. (2,610 kg.)

DIMENSIONS

Length	65.6 ft. (20 meters)
Height	20.8 ft. (6.4 meters)
Wingspan	64.9 ft. (19.8 meters)
Wing area	424.5 sq. ft. (39.4 sq. meters)
Wing aspect ratio	9.923
Cabin length	30.8 ft. (9.4 meters)
Cabin width	6.9 ft. (2.1 meters)
Cabin height	5.8 ft. (1.8 meters)
Standard cabin configuration	30 passengers

PERFORMANCE

Takeoff field length requirement at 23,810 lb. with standard conditions and zero wind	4,530 ft. (1,380 meters)
Landing field length requirement at 23,260 lb. with standard conditions and zero wind, Part 135 requirements	4,460 ft. (1,360 meters)
Rate of climb on one engine at 23,810 lb.	460 fpm.
Maximum cruise true airspeed at 23,380 lb. and at 20,000 ft.	294 kt.
Maximum range with 30 passengers with a fuel reserve for 100 naut. mi. diversion and 45 min. holding	515 naut. mi.
Range with maximum usable fuel and corresponding payload to reach 23,810 lb. with same fuel reserves	1,560 naut. mi.

DHC-8 meets goals for performance

David M. North/Downsview, Ontario
April 23, 1984

De Havilland Aircraft of Canada's DHC-8 (Dash 8) exhibits excellent handling characteristics and performance that matches goals the company set for the 36-passenger commuter/ corporate aircraft at the beginning of its development program in 1980.

The DHC-8 is the first of the new generation of 30-50-passenger commuter aircraft to be flown by this *Aviation Week & Space Technology* pilot. Other aircraft scheduled to enter commuter operations within the next few years include the Saab-Scania/Fairchild Industries 340, Embraer EMB-120 Brasilia, CASA/Nurtanio CN-235 and the Aeritalia/Aerospatiale ATR42. The Short Brothers SD 330 and SD 360 are already in U.S. commuter service, and the modified Fokker 50 is expected to enter service in the mid-1980s.

The No. 1 DHC-8 prototype, with Canadian registration C-GDNK, was flown from de Havilland's headquarters here with Adam W. Saunders, director of flight operations. The first leg of the evaluation flight from the Downsview Airport to the London, Ontario, Airport about 90 naut. mi. away would follow a commuter-type night profile. After landing and takeoff at London, the airwork portion of the flight would be conducted on the return leg to Downsview.

De Havilland aircraft of Canada DHC-8 prototypes have accumulated more than 700 hr. in flight test since the aircraft first flew in June, 1983. The four flight-test aircraft will be joined this summer by a fifth aircraft, which will be used for function and reliability testing. The DHC-8, powered by two Pratt & Whitney of Canada PW120 turboprop engines with a maximum rating of 2,000 shp., is scheduled to be certificated in September. The No. 1 prototype, with Canadian registration C-GDNK, was the aircraft flown in the evaluation flight.

Saunders described a few of the features of the Dash 8 during preflight. The single-point refueling receptacle is located in the rear of the right engine nacelle. The commuter configuration has a fuel capacity of 825 U.S. gal., and the extended range corporate version has nearly twice that capacity. A 50-×-60-in. baggage door on the rear fuselage is designed so that the aircraft can be used as a cargo carrier. The aircraft has a movable rear bulkhead to accommodate passengers and cargo in a variety of configurations.

The interior of the No. 1 prototype still contained flight test instrumentation for monitoring aerodynamic performance. This capability, covering more than 30 flight parameters, eased my task of recording time, speeds, fuel flow and other flight characteristics. The flight test engineer for the flight was Waldemar Krolak.

Cockpit instrumentation in the No. 1 prototype is similar to that installed on a standard commuter aircraft, except for flight test instruments, such as sideslip, angle of attack and calibrated airspeeds. Inputs to these instruments came from a test boom located on the right wing. The aircraft I flew did not have the optional Sperry electronic

Standard commuter interior for the de Havilland DHC-8 contains nine rows of seats in a four-abreast arrangement. The UOP Aerospace fixed-back seats provide a 31-in. pitch and allow a minimum aisle width of 15 in. between armrests. The standard interior includes a wardrobe behind the cockpit and in front of the main door on the aircraft's left side and a lavatory and buffet across from the airstair door. The approximately 300-cu.-ft. baggage compartment in the rear of the aircraft's fuselage is reached through a 50-in.-×-60-in. cargo door.

flight instrument system. From my position in the left seat, the Dash 8 instrument panel was uncluttered and there was ample room for additional instruments if required. The gross weight of C-GDNK was calculated by Krolak to be 29,972 lb., including 4,000 lb. of fuel. Also included in the total weight was water ballast used to achieve close to a 25% center of gravity. The actual center of gravity was 24.6%.

De Havilland originally had planned to certificate the DHC-8 at a maximum takeoff weight of 30,500 lb., and the commuter version was to have an operational empty weight of 20,176 lb. The targeted empty weight was exceeded by approximately 1,400 lb. during aircraft development. With an increase to a maximum takeoff weight of 33,000 lb., de Havilland has been able to absorb the approximately 5% weight increase and provide another 1,100 lb. for payload. The takeoff gross weight for my flight was approximately 91% of the maximum takeoff weight.

Automatic sequence

Both Pratt & Whitney PW120 turboprop engines were started by an automatic sequence, with internal turbine temperatures staying well below the maximum limits on the 13°C (58°F) day. The fuel flow at idle power was 255 lb./hr./engine.

A total of 60 lb. of fuel was used from engine start to the beginning of takeoff roll, 14 min. later. Steering during taxi is accomplished with a handle to the left of the pilot. A steering engage button is on top of the handle. The handle allows the nose wheel to travel up to 60 deg. from the centerline. For small turns and during takeoff the rudder pedals control up to 7 deg. of nose wheel travel. There was no nose wheel shimmy during the taxi to Runway 15 or during the takeoff.

I applied takeoff power to the engines, achieving a maximum of 90% torque. An automatic power reserve feature in the Dash 8's turboprops allows either engine to achieve 100% torque, or 2,000 shp., if the other engine should fail at takeoff. The normal takeoff rating for the PW120s on the Dash 8 is 1,800 shp.

Rotation speed

The aircraft was rotated to a 10-deg. nose-up attitude at the rotation speed of 81 kt. The V_1 takeoff decision speed also was 81 kt., and the V_2 takeoff safety speed was calculated to be 89 kt. Takeoff roll was less than 1,400 ft. With winds at 130 deg. and 10 kt., there was a 9-kt. headwind component. Fuel flow during the takeoff roll was 930 lb./hr./engine.

Less than 1 min. after the wheels left the runway, the Dash 8 was climbing through 2,000 ft. at a speed of 140 kt. Fuel flow at 2,000 ft.

was 770 lb./hr./engine. Another 2 min. was required to reach 5,000 ft. at a climb rate of 1,500 fpm and the same 140-kt. climb speed. The 140-kt. climb speed resulted in a relatively flat aircraft attitude and afforded excellent visibility from the left seat of the surrounding airspace and the ground. The wrap-around windshield provided this good visibility throughout the flight. Another benefit of the smooth nose configuration was little noise in the cockpit, especially during the descents made near the aircraft's maximum speed. Saunders concurred that the cockpit noise was much less than that experienced in the company's DHC-7, which I first flew in 1977 (AW&ST June 13, 1977, p. 40).

As the aircraft passed through 10,000 ft. 6 min. after takeoff, the speed was still 140 kt. and the rate of climb had decreased to 1,300 fpm. Fuel flow at the 10,000-ft. mark was 708 lb./hr./engine. The assigned cruising altitude of 12,000 ft. was reached 7 min. 21 sec. after liftoff at Downsview. A total of 240 lb. of fuel had been used from engine start, of which 60 lb. had been used during taxi.

A maximum cruise torque setting of 80% was selected at 12,000 ft. At the static –8°C temperature, the indicated airspeed was 217 kt. and the true airspeed was 257 kt. De Havilland claims a maximum true airspeed for the Dash 8 of 270 kt. at a 30,500-lb. gross weight and 15,000-ft. altitude. Fuel flow at the maximum cruise power setting was 680 lb./hr./engine.

Long-range cruise speed for the Dash 8 at 12,000 ft. was determined to be 203 kt. I set the power at 45% torque, which resulted in an indicated airspeed of 176 kt. and a fuel flow of 475 lb./hr./engine. No unnecessary turns or maneuvers were made during the flight to the London Airport so that the performance figures would approximate those of an aircraft in normal commuter operations.

A descent was made approaching London Airport at 170 kt. and idle power. The fuel flow during the descent was 240 lb./hr./engine. The approach to the glideslope at London Airport was flown at 140 kt., gradually slowing to 110 kt. and then 92 kt. with the landing gear down and the flaps at the 35-deg. setting. The power required to maintain the glideslope to Runway 15 at London was 20% torque at a fuel flow of 500 lb./hr./engine. I found the Dash 8 extremely stable and responsive to small inputs from the flight controls during the approach.

Touchdown at the London Airport was 32 min. after liftoff from Downsview. The landing was firm, but not hard, as I searched for the correct eye-to-landing-gear reference point after initially leveling off at a higher altitude.

The distance between the two airports is approximately 90 naut. mi. Allowing for taxi time and possible delays in landing and takeoff, a block-to-block time of 45 min. would be reasonable. Total fuel used was 890 lb.

De Havilland expects operators to be able to dispatch the Dash 8 with the electronic fuel controls on the engines inoperative. Saunders said the specific fuel consumption of the preproduction engines and an earlier fuel control configuration in the No. 1 prototype is higher than it will be in aircraft with newer engines and a fuel control with a different schedule.

The takeoff from London Airport was a repeat of the earlier one from Downsview. Although the Dash 8 has not been labeled a short takeoff and landing aircraft by de Havilland, the aircraft is easily capable of operating from a 3,000-ft. runway in standard atmospheric conditions.

Passing through 14,000 ft. 14 min. after takeoff from London Airport, the rate of climb was 1,400 fpm, and the fuel flow was indicating 660 lb./hr./engine. Another 2 min. were required to reach our cruising altitude of 17,000 ft.

At cruising altitude, a maximum cruise power setting of 75% torque yielded a 206-kt. indicated airspeed and a 262-kt. true airspeed. At the slightly warmer than standard day, the fuel flow was 630 lb./hr./engine at a setting of 1,050 rpm. The fuel flow dropped to 430 lb./hr./engine at a long-range cruise torque power setting of 43%. True airspeed at the long-range setting was 207 kt.

A 5.2-psi. differential at 17,000 ft. equaled a 3,600-ft. cabin altitude for the DHC-8. The maximum pressure differential is 5.5 psi. During the entire flight, including stalls and simulated engine-out operation, there was no indication of pressurization surges or bumps in the cockpit.

While still at 17,000 ft. and with a block of altitude to 13,000 ft. assigned by air traffic control, I performed numerous level and varying altitude steep turns. The Dash 8 was very responsive in pitch control, but less so in roll control, especially at the faster speeds. Saunders said there was some friction in the aileron control on the No. 1 prototype, but that it was being corrected and de Havilland was attempting to lighten roll control by minor adjustments in the control system.

The aircraft's roll control at speeds near 100 kt. was more than adequate. At these slower speeds, roll control is aided by hydraulically powered, high-aspect-ratio spoilers on the trailing edge of the wing shroud. Saunders said that even with the friction in the aileron system, the roll control was more responsive and positive than that of the four-engine DHC-7.

The Dash 8 does not have an electrical elevator trim switch on top of the yoke, but the number of times I had to use the manual wheel trim system was minimal. The aircraft is very stable in all flight regimes, and I found myself releasing the yoke to see whether the aircraft was in trim or whether I was overcontrolling the aircraft. During landing gear and flap deployment there was little pitch change in the aircraft.

During the development program, de Havilland modified the wing between the engine nacelles and the fuselage with a drooped leading edge. The company has determined in flight testing that the leading edge modification has increased the maximum lift of the aircraft and reduced stall speed. This has enabled the Dash 8 to meet its forecast airfield performance at the increased 2,500 lb. maximum takeoff gross weight.

In clean configuration and with idle power at 15,000 ft., the Dash 8's stick shaker activated at 89 kt. and the aircraft stalled at 83 kt. The angle of attack prior to the stall was 26 deg., and Krolak said that there was 30 lb. of elevator pull in the stall condition. The stall was straightforward and the aircraft showed no tendency to fall off on either wing. Power response for stall recovery was quick and positive.

I lowered the flaps to 15 deg., and with the power at idle, the stick shaker started at 75 kt. and the aircraft stalled at 71 kt. at a 31-deg. angle of attack. With partial engine power, the aircraft stalled at 65 kt. after the shaker activated 3 kt. faster. Angle of attack reached during this maneuver was 24 deg.

The Dash 8's shaker started at 71 kt. and the aircraft stalled at 66 kt. and a 31-deg. angle of attack with the flaps in the 35-deg. landing configuration and the power at idle. In the same configuration, but with partial power, the stick shaker activated at 65 kt. and the aircraft stalled at 62 kt. at a 28-deg. angle of attack.

To simulate a right engine failure, power on that engine was reduced to 15% torque and the left engine was advanced to the maximum allowable at 14,000 ft. The amount of rudder trim required to compensate for the engine-out situation was less than half that which was available. Maintaining a speed of 140 kt. with constant power resulted in a climb of 500 fpm.

The descent to the landing approach at Downsview was made at slightly less than the maximum speed of the aircraft. At 12,000 ft., the speed was 240 kt. indicated and, the fuel flow was 755 lb./hr./engine. A speed of 110 kt. was used on the approach to Runway 15 at Downsview. During the long straight-in approach, the engine power was set to maintain 110 kt., and when flaps and landing gear extended, the same power setting provided a comfortable margin above the 86-kt. reference speed. Once established on a visual glide slope at a rate of descent of 500 fpm, and with the left wing down to compensate for an increased crosswind since takeoff, the Dash 8 could almost have been flown hands off. The No. 1 prototype does not have an autopilot.

My landing at Downsview was better than the one at London Airport. The left landing gear touched down smoothly followed by the right landing gear and nose wheel. With reverse pitch on the Hamilton Standard propellers but little use of the brakes, the Dash 8 was slowed to taxi speed in less than 2,000 ft.

Total elapsed time from engine start to engine shutdown was 2 hr. 1 min., and the airborne time, excluding 4 min. on the runway at London Airport, was 1 hr. 43 min. The total fuel used for the flight was 2,200 lb., with 1,800 lb. remaining at the de Havilland ramp.

There is little airframe commonality between the DHC-7 and the DHC-8, R.A. Paul Stafford, Dash 8 project engineer, said. The similarities are in the aircraft systems, although the Dash 8 has a different flap drive system and a single air conditioning system compared with the dual system on the DHC-7. The fuselage diameter of the Dash 8 is 4 in. less than the Dash 7.

"While the fuselage and wing of the Dash 8 are different from those of the Dash 7," Stafford said, "we did not experience any surprises in either the structural or flight characteristics portion of the development program. Our entire program on the Dash 8 followed along the lines of the Dash 7."

The No. 2 DHC-8 prototype is flying from Marana Airpark in Arizona to expand the performance figures for hot temperatures and high altitude. The No. 2 aircraft also has been assigned aircraft systems test. Dash 8 No. 3 has accumulated more than 200 hr. and is being used to complete the icing tests, antenna patterns testing and evaluation of exterior noise levels.

Cold weather and engine assessment testing plus the evaluation of the Dash 8's avionics systems is being conducted with the No. 4 prototype. The function and reliability portion of the certification testing will be done on the No. 5 prototype, scheduled to fly for the first time in June and be the first aircraft with a full commuter interior.

Damage tolerance testing of the airframe is now underway at the company's manufacturing facility here. Detail fatigue testing of the aircraft is scheduled to begin this summer. Ultimate design load testing of the wing and pressure testing of the aircraft's forward fuselage have been completed.

De Havilland expects to receive Canadian type approval in September, followed closely by Federal Aviation Administration certification. The No. 6 aircraft is scheduled to be the first delivered to a regional operator, either in September or October. NorOntair is to receive the first one.

Within a year after the beginning of Dash 8 deliveries, the company expects to deliver three of the five prototype aircraft, with full interiors, to operators. The No. 1 prototype will be retained as a test aircraft, and No. 4 will be used as a demonstrator. De Havilland expects to deliver 15 Dash 8s by May, 1985, and to be up to a production rate of four monthly in 1985. The company has sufficient tooling to build six aircraft a month.

The first corporate Dash 8 is scheduled to be delivered to Innotech Aviation, the agent for the aircraft in Canada and the U.S., in January, 1985. Innotech expects to deliver the only corporate Dash 8 now on order in the summer of 1985. The corporate version has as standard equipment a Turbomach Solar auxiliary power unit, Sperry electronic flight instrument system, extended-range fuel tanks and a custom interior to match the needs of the operator. The price of the corporate DHC-8 is approximately $6.1 million (U.S.).

Price of the commuter version of the Dash 8 is approximately $5.5 million in U.S. dollars. Among U.S. regional airlines with orders for the Dash 8 are Henson Airlines with eight, Rio Airways with six, Rocky Mountain Airways with six and Southern Jersey Airways with four. Canada's Air Atonabee expects to receive five Dash 8s, and Time Air, also of Canada, has orders for four.

DHC-8 SPECIFICATIONS

POWERPLANT

Two Pratt & Whitney of Canada PW120 turboprop engines with a normal takeoff rating of 1,800 shp. An automatic power reserve feature enables each engine to be operated at its maximum 2,000-shp. rating with the failure of one engine.

PROPELLERS

Two Hamilton Standard 14SF four-blade, 13-ft.-dia., constant speed with full feather and full reversible capabilities.

WEIGHTS

Maximum takeoff weight	33,000 lb. (14,969 kg.)
Maximum landing weight	32,400 lb. (14,696 kg.)
Maximum zero fuel weight	31,000 lb. (14,061 kg.)
Operational empty weight	21,590 lb. (9,793 kg.)
Maximum payload	9,410 lb. (4,268 kg.)
Usable fuel (commuter)	5,490 lb. (2,490 kg.)
Usable fuel (corporate)	10,312 lb. (4,677 kg.)

DIMENSIONS

Length	73 ft. (22.3 meters)
Height	24.6 ft. (7.5 meters)
Wingspan	85 ft. (25.9 meters)
Wing area	585 sq. ft. (54.4 sq. meters)
Wing aspect ratio	12.0
Cabin length	30.2 ft. (9.2 meters)
Cabin width	8.2 ft. (2.5 meters)
Cabin height	6.2 ft. (1.9 meters)
Standard cabin configuration	36 passengers

PERFORMANCE

Takeoff field length requirement at 33,000 lb., and with flaps at 15 deg.	2,820 ft. (860 meters)
Landing field length requirement at 32,000 lb. and with flaps at 35 deg.	2,990 ft. (911 meters)
Maximum cruise true air speed at 30,500 lb.	270 kt.
Maximum range with 36 passengers, instrument flight fuel reserves at maximum cruise thrust	900 naut. mi.
Maximum range with 36 passengers, instrument flight fuel reserves at long-range cruise thrust	950 naut. mi.
Maximum range with extended range fuel tanks	2,650 naut. mi.

Chapter 4

Small turboprops

CBA-123 turboprop has jet-like qualities

David M. North/Sao Jose dos Campos, Brazil
September 7, 1992

Embraer hopes to outwit recessing and sluggish commercial market by continuing certification tests while negotiating for sale of military version.

The Embraer CBA-123 exhibits the performance qualities desired by commuter operators, but market conditions and a relatively high selling price of the 19-passenger turboprop have combined to delay series production. The development of the twin-turboprop was begun in the mid-1980s, when the market outlook for a regional transport of this type was relatively good. Since that time, the worldwide recession and downturn in aircraft purchases have forced the largest South American aircraft manufacturer to rein in its plans for the CBA-123 Vector.

While the Brazilian company is conducting flight testing of the two CBA-123 prototypes, a go-ahead decision on the commercial version is on hold. To give itself time for the market to recover and yet continue development and aircraft certification, Embraer is negotiating with both the Brazilian and Argentinian governments for CBA-123 deliveries to military units within the countries. Under the agreement, if signed, Brazil would receive 40 and Argentina 20 Vectors.

This *Aviation Week & Space Technology* pilot had the opportunity to fly the No. 1 prototype, PT-ZVE, during a recent trip to Embraer's headquarters here. While there, I noticed the lack of any work being done on the 19-passenger pressurized transport. Jigs and tooling for the production aircraft were in storage in one of the hangars.

No. 1 CBA-123 prototype was flown by Managing Editor David North during a 1.3-hr. evaluation.

"When we launched the CBA-123 program in 1985 under the government umbrella, the commuter situation was different," Ozires Silva, president of Embraer, said. "Now there are a number of aircraft this size sitting on the ground. We are not shutting the door on the aircraft, but have submitted the program to both governments to see whether we will receive any support."

Embraer signed an agreement with Argentina's Fabrica Militar de Aviones in 1987 for joint CBA-123 development. At that time, Embraer was to have 70% of the work and FMA the remaining 30%. The FMA share was later reduced to 20%. The third Vector is being built at the FMA facility in Cordoba, Argentina, and is scheduled to fly by the end of the year. This is a slip of two years in the first flight of the third prototype from the schedule set in mid-1990.

About $180 million has been spent on the program for research, development and flight testing. Another $120 million would be required to put the aircraft into series production and deliver the first unit, according to Horacio Aragones Forjaz, Embraer's technical director.

More than 95% of the CBA-123 performance requirements have been completed, Forjaz said. At least 40% of the total certification requirements have been met, while 72% of the structural tests have been completed. Embraer has worked continually with the U.S. Federal Aviation Administration, and the target for CBA-123 U.S. type certification is 1993.

"Garrett has been providing full support for the engines, and Collins for the avionics," Forjaz said. "Because of production commonality with the [EMB120] Brasilia, and the work already completed, we could deliver CBA123s within nine months to a year after getting the go ahead."

Forjaz defended the relatively high initial price of the CBA-123 by stressing its performance and the predicted lower operating costs. Although he was reluctant to place a set price on the CBA-123 because of the many uncertainties, he said $6 million was possible. Embraer has embarked on a cost reduction program to lower the acquisition price by some $400,000. The cut would come from simplified aircraft systems, such as a single hydraulic system to the rudder, and by production efficiencies.

The No. 1 prototype CBA-123 I was to fly did not have any of the cost modifications incorporated, but it still had the flight test equipment. Its first flight was on July 18, 1990, and 457 hr. had been logged prior to my evaluation. The second prototype first flew on Mar, 15, 1991. Embraer has logged more than 850 hr. on the two aircraft.

Embraer chief test pilot Gilberto Pedrosa Schittini accompanied me on the flight from Sao Jose dos Campos airport. I also flew with him in the proof-of-concept Tucano H (AW&ST Apr. 20, 1992 p. 46). Following the walk-around, I took the left seat and Schittini occupied the right seat. One observer and Embraer flight test engineer Luiz Fernando Tedeschi Oliveira also came along.

The ramp weight of the CBA-123 was 20,967-lb. (9,509 kg.), or only slightly above the 20,944-lb. maximum takeoff weight. Embraer had initially planned for a 19,841-lb. (9,000-kg.) maximum takeoff weight, but increased it partly to compensate for a higher empty weight while keeping the same planned payload. Included in the ramp weight was 3,528-lb. of fuel, 617-lb. of water ballast, 1,320-lb. of flight test equipment and 700 lb. allocated for crew and passengers. Maximum usable fuel capacity is in the 4,950-lb. range.

The flight deck was almost spartan in appearance because of the relatively large instrument panel, the lack of many small instruments and the roominess of the cockpit. The efforts by Embraer—as well as Schittini and his flight crew—to reduce pilot workload and incorporate human factors in the cockpit design were evident. The overhead panel contained the systems switches and indicators, and was designed to have all lights out when in flight and everything working normally.

Collins Pro Line digital avionics are used. This standard package includes the Collins engine indication and crew alert system (EICAS). An all cathode-ray-tube display is used with some basic backup instruments installed on the center instrument panel. Standard checklists have yet to be incorporated on the displays, but they are to be added later, Schittini said. Maintenance messages also will be added but will not be displayed below 10,000 ft., according to FAA regulations.

Both Garrett engines were started by use of switches on the overhead panel. There was little change in the level of cockpit noise to in-

Embraer Vector is powered by two turboprop Garrett engines in a pusher propeller configuration.

CBA-123 is equipped with Collins Pro Line digital avionics. Uncluttered cockpit provides good situational awareness.

dicate that both engines were running. Their quietness was a factor throughout the flight. When I expected an aural cue to a power change, there was none available. The counter rotating propellers are located almost 12 ft. behind the rear-most seat in the passenger cabin.

The nose-wheel steering handle was used to taxi the aircraft to the active runway. The hydraulically powered steering gives wheel deflection of 50 deg. either side of center, while the rudder pedals provide 7 deg. from center. The brakes were very effective and smooth during the taxi, and had no tendency to grab.

I aligned the aircraft to the runway centerline and added power to a 98% torque setting. Brakes were held partway through the power advancement. All the takeoff target speeds had been calculated to be 112 kt. I was slightly late in rotating the nose, so actual rotation speed was 117 kt. The takeoff distance was 3,940 ft. on the 29C (84F) day. The gear and flaps were retracted shortly after takeoff, and a 170 kt. climb speed was established. One minute was required to reach 4,000 ft. after brake release from the 1000-ft.-msl. airport. A total of 4.6 min. from brake release was required to reach 10,000 ft., while the rate of climb passing through 10,000 ft. was 1,700 fpm. The altitude of 15,000 ft. was reached in 8 min.; the rate of climb was 1,400 fpm. It required

12 min. from brake release to reach 20,000 ft., and the rate of climb was 900 fpm. The cruising altitude of 23,000 ft. was reached in 16 min. The total fuel used was 420-lb. from engine start, and the aircraft's gross weight was near 20,530-lb.

The autopilot had been engaged at 16,000 ft. to observe the transition from climb to level flight. The CBA-123 leveled off smoothly at 23,000 ft., and engine torque remained high for 3 min. until cruise speed was attained. With the torque set at 79% and propeller rpm at 91%, indicated airspeed was 207 kt. At the ISA +14C day, true airspeed was 304 kt. Fuel flow at the cruise speed was 860-lb./hr.

Cabin pressurization was smooth during the climb and throughout the flight. The digital electro-pneumatic control system is designed to maintain a maximum 8.2 psi. cabin/ambient differential pressure. The air-conditioning system also was effective during the flight and is supplied by two air cycle machines.

The aircraft was fitted with one early prototype Garrett engine on the right side and a production TPF351-20 engine on the left. Schittini said that speeds and fuel flows would be slightly better when production engines were mounted on both sides. The performance of the CBA-123 also would be improved when Embraer finished the modifications to the full authority digital electronic control (FADEC). Schittini said the changes were still being added.

Aside from the planned performance improvements, actual maximum cruise speeds are still below those predicted for the twin-turboprop prior to flight test. Embraer had predicted close to a 350-kt. true airspeed at 23,000 ft. on a standard day with a 17,600-lb. aircraft.

Embraer installed winglets on the No. 1 prototype early in the development phase, but they were not incorporated when flight data showed no significant increase in performance. The current wing has proved to be efficient in all flight regimes.

Flight tests have verified a 328-kt. maximum cruise speed at the same parameters. The disparity between forecast and achieved maximum cruise speeds at 10,000-20,000 ft. is not as large as in the 20,000-30,000-ft. range. Other predictions, such as climb performance, were more closely met or improved on during flight test.

I found that pitch control was responsive at near the maximum cruise speeds, but roll control forces were higher than expected. Schittini said that initially Embraer had incorporated a camber in the ailerons that resulted in lower forces for roll control at the higher speeds. The camber was eliminated to optimize control forces over the CBA-123's entire speed range. Roll control forces were better than expected and very responsive at the lower speeds.

In the 20,000-ft. altitude range, throttles were retarded to idle while in a clean configuration. I trimmed the aircraft to 130 kt. The stick shaker activated prior to the stick pusher forcing the control yoke forward at 99 kt. The alpha rate was slightly higher than 1 kt./sec. entering the stall. The angle of attack at the stick pusher was close to 22 units. The stick pusher was installed because the Vector would usually stall to the right and far exceed the FAA guidelines for wing drop during stalls. Embraer plans to certify the Vector to Federal Aviation Regulations Part 25.

The landing gear and flaps were then extended to evaluate slow flight. Roll control was much crisper in the 120-140-kt. range, and pitch control was precise. With the throttles at idle, the CBA-123 was trimmed to 111 kt. The stick pusher activated at 86 kt. with an AOA of 21 units. The aircraft again had a tendency to roll to the right during recovery.

At 17,000 ft., the torque was set at 95% and rpm to 91%. The speed attained in level maximum cruise was 233 kt., or a 304 kt. true airspeed. Fuel flow was 1,060-lb./hr. (480 kg./hr.).

Two Vector prototypes are in flight test. The third, produced by FMA in Argentina, is due to fly by the end of the year.

I reduced power and descended to 12,000 ft. to fly a simulated engine-out maneuver. The thrust lever for the left engine was retarded to a 10% torque setting. The right engine thrust lever was set to its maximum 92% torque. Maintaining 110 kt., I was able to achieve a 300-fpm climb rate. There was ample right rudder available to compensate for yaw. Schittini said that through the rudder electrical actuation system and the 19.3-sq.ft. rudder, there is more than adequate control authority available down to the lower speeds.

After this sequence, we headed back to Sao Jose dos Campos airport. An instrument landing approach was flown to the active runway. I found that the Collins Pro-Line EFIS-86 display made situational awareness easy to maintain.

Reaffirming my earlier observation and after flying the Vector, I found the entire cockpit display had been well thought out: human engineering was a key factor in its design. Schittini said the Collins/Embraer team listened to his and other Embraer pilots' input during the design phase.

The ILS was flown at 120 kt. with landing gear and flaps at 25 deg. The CBA-123 was very responsive to pitch and roll inputs, with no tendency to over control. The speed was 110 kt. over the end of the runway, and touchdown with a slight flare occurred at 105 kt. The touchdown was smooth on the French ERAM manufactured landing gear.

I performed a touch-and-go with a left downwind for a final landing. The visibility from the left seat was very good, and I was able to spot other aircraft easily in the pattern. The second landing was a repeat of the first.

The CBA-123 handles much like a jet in flight and in the landing pattern, and with the absence of any engine noise in the cockpit, the similarity is heightened.

Total fuel burn for the 1.1-hr. flight and 1.3-hr. blocks-to-blocks time was 970 lb. (440 kg.). Schittini said that a rough estimate of fuel burn for a long-range flight is 900-lb. in the first hour and 800-lb. for the remaining flight hours.

Embraer has built a high-technology, pressurized regional aircraft with performance, cabin amenities and quietness that exceed others in its class. The question remains whether regional airlines—many with financial problems and with an overabundance of similar size aircraft—will be able to afford the Vector.

If Embraer is able to establish the Vector in the regional aircraft market with a base production run, the CBA-123 has many of the attributes that could be attractive to the corporate operator.

EMBRAER CBA-123 SPECIFICATIONS

POWERPLANT

Two Garrett TPF351-20 engines of 1,300-shp. each driving Hartzell six-bladed, pusher, counter-rotating propellers.

WEIGHT

Maximum ramp weight	21,032 lb. (9,540 kg.)
Maximum takeoff weight	20,944 lb. (9,500 kg.)
Maximum landing weight	20,392 lb. (9,250 kg.)
Maximum zero fuel weight	18,739 lb. (8,500 kg.)
Basic operating weight	13,988 lb. (6,345 kg.)
Maximum usable fuel	733 U.S. gal. (2,775 liters)

DIMENSIONS

Length	59.4 ft. (18.1 meters)
Wingspan	58.2 ft. (17.7 meters)
Height	19.6 ft. (6.0 meters)
Cabin length	21.2 ft. (6.5 meters)
Cabin width	6.9 ft. (2.1 meters)
Cabin height	5.8 ft. (1.8 meters)

Comfort, performance mark larger 1900D

Edward H. Phillips/Wichita
August 31, 1992

In the highly competitive regional airline market, Beech has a winner. The stretched aircraft has docile, predictable handling and well-harmonized controls.

The Beechcraft Model 1900D's stand-up cabin, upgraded cockpit and improved airframe systems provide regional airline operators with increased performance as well as a higher level of cabin comfort for passengers.

This *Aviation Week & Space Technology* pilot flew the 12th production 1900D on a 1-hr. 30-min. flight from Beech Field in Wichita to Forbes Field near Topeka, Kan. I was accompanied by Dennis G. Hildreth, engineering flight test pilot, and Joseph A. Papke, manager, international airline sales for Beech.

Beechcraft's advanced, 19-seat 1900D has evolved from the company's successful Model 1900C regional transport that was introduced in 1984. Production ceased in 1991 after 255 1900C aircraft were delivered to 17 U.S. and seven international airlines, according to Papke.

During the preflight inspection, Hildreth pointed out the aircraft's larger stabilons and strakes under the aft fuselage. Although the 1900C featured stabilons, the strakes are unique to the 1900D. The stabilons and strakes provide increased longitudinal stability and eliminate the

Each of the 1900D's Pratt & Whitney Canada PT6A-67D turboprop engines is flat-rated to 1,279 shp. The aircraft is easy to fly and possesses good handling characteristics typical of Beechcraft's turboprop King Air family.

need for a stability augmentation system to handle the aircraft's wide center of gravity range, he said.

The new aircraft's dual tailets also are larger than the 1900C's. Mounted under the horizontal stabilizer, the tailets provide additional directional stability. Composite winglets similar to those installed on the Super King Air 350 have replaced conventional wingtips used on the 1900C, and the 1900D retains the large, aft cargo door found on the previous version. It also shares the 1900C's 665.4 gal. usable fuel capacity carried within the integral fuel cells in the wings.

Entering the cabin, Papke explained that the 1900D is a much more mature, fully developed light transport than the 1900C, which was essentially a stretched cabin version of the Super King Air B200 business aircraft. In creating the 1900D, Beech engineers concentrated on improving passenger comfort and easing cabin entry.

To achieve those goals, the aircraft has a significantly larger interior that boasts a 71-in. height from the flat floor to the ceiling—14 in. higher than its predecessor. in addition, cabin volume has been increased 28.5% to 640 cu. ft. from 498 cu. ft.

Based on passenger reaction, Beech has apparently succeeded in its quest to improve both regional airline and customer acceptance of the 1900D. Farmington, N. M.-based Mesa Airlines, operating as a United Express affiliate, has ordered 58 1900Ds and currently has about 10 in service. Passengers comment that "they really have appreciated the difference" in the 1900D's larger cabin, George Lippemeier, director of flight operations, said.

"There's a lot more room in this aircraft and it seems easier to move around in. I don't have to take off my hat to get in and I can leave it on while I'm in my seat," one Mesa customer said.

To further improve passenger comfort, Beech has installed larger seats similar to those used in the McDonnell Douglas MD-80 and Boeing 727/737 transports. Seat pitch has been increased to 30 in., and each cushion is designed to act as a flotation device.

In addition to paying close attention to passenger needs, Beech has not neglected the flight crew. Although the cockpit configuration is essentially the same as the 1900C, basic avionic and flight instrumentation has been upgraded to include a Collins Pro-Line EFIS-84 electronic flight instrument system with 4-×-4-in. color cathode ray tube displays for the captain and first officer.

Each pilot's panel has an electronic attitude director instrument and an electronic horizontal situation instrument with reversionary mode capability between upper and lower displays. Information from the Collins WXR-350, four-color weather radar is displayed on the EHSI-84 CRT on either panel. A digital flight data recorder and cockpit voice recorder also are standard.

The Beech 1900D has a 71-in. height from the flat floor to the ceiling. The 19-seat aircraft has a cabin volume of 640 cu. ft.

I strapped into the left seat and Hildreth did the same in the first officer's chair. Starting the two Pratt & Whitney Canada PT6A-67D turboprop engines was straightforward. Each powerplant is flat-rated to 1,279 shp. and turns a four-blade, fully reversible Hartzell propeller.

Beechcraft 1900D features winglets to improve high density altitude performance as well as tail-mounted stabilons and tailets to eliminate the need for a stability augmentation system. Large, ventral-mounted strakes improve longitudinal stability.

In preparation for taxiing to Runway 36 at Beech Field, I selected the TAXI mode of the power steering system that operates via the rudder pedals. The TAXI mode provided excellent directional control. A PARK mode further increases steering authority for taxiing in crowded ramp areas with one engine shutdown. Visibility through the cockpit windows was more than adequate.

The aircraft weighed 15,400 lb. at takeoff—nearly 1,500 lb. below the aircraft's maximum takeoff weight of 16,950 lb. Using the TAXI mode, I aligned the nosewheel with the runway centerline and then disengaged the power steering system. Flaps were set to 17-deg. deflection.

I held the brakes and advanced the power levers to a takeoff power setting of 3,700 ft. lb. of torque. After brake release, Hildreth monitored engine power as the aircraft quickly accelerated to the calculated rotation speed (V_r) of 100 KIAS. I rotated, establishing an airspeed of 160 KIAS and a positive climb rate before retracting the landing gear and flaps.

During the climb, I reduced propeller speed (N_p) to 1,550 rpm and reset torque to 3,750 ft. lb. en route to Forbes Field at Flight Level 230. Initial rate of climb was 2,300 ft./min.

The cockpit noise level was not uncomfortably high, and the 1900D handled the turbulence well as it climbed through a mass of Kansas-bred summer cumulus clouds. The aircraft made no special demands of me, but the control forces in roll and pitch are understandably heavier than those of its Model B200 and Model 350 Super King Air siblings.

The environmental and pressurization systems maintained a comfortable temperature and cabin pressure throughout the flight. For ground cooling in summer months, a vapor-cycle system rated at 39,000 BTUs is installed and uses an engine-driven compressor mounted on the No. 2 powerplant accessory section. An air cycle machine capable of 46,000 BTUs handles cooling demands in flight, and dual bleed air systems rated at 64,000 BTUs provide heat.

I liked the Collins EFIS-84 displays. They were easy to see and interpret and represent a major improvement compared with the 1900C's conventional electro-mechanical instrumentation. Passing through 10,000 ft., I transitioned to a climb airspeed of 140 KIAS up to FL230.

Upon reaching the assigned cruising altitude, I set power to 96% gas generator speed N_g and 2,600 ft. lb. of torque, and reduced N_p to 1,550 rpm. The aircraft accelerated to 193 KIAS. I then reduced N_p to an intermediate cruise setting of 1400 rpm and torque to 2,200 ft. lb. The 1900D indicated 173 KIAS and 250 KTAS while each engine burned 330 lb. of fuel per hour.

Cockpit upgrades include a Collins EFIS-84, four-tube electronic flight instrument system with dual displays.

With 19 passengers on board, at an aircraft weight of 10,550 lb. and a high-speed cruise power setting, the aircraft can fly 705 naut. mi. at its maximum certified altitude of 25,000 ft. At 12,000 ft. range decreases to 518 naut. mi. and falls to 481 naut. mi. at a cruising altitude of 8,000 ft. Maximum cruise speed at 15,000 lb. at 25,000 ft. is 278 KTAS.

After cruising for about 20 min., Kansas City Center cleared us for a descent into Forbes. I reduced power and established a 1,500 ft./min. descent rate into the terminal area as ATC vectored me for a visual approach to Runway 13.

With the landing checklist completed, Hildreth calculated a reference speed (V_{ref}) of 113 KIAS for the approach. I entered a left downwind, extended the landing gear and 17 deg. approach flaps below 180 KIAS and retrimmed the aircraft.

On final to Runway 13, flaps were deflected to 35 deg. for landing and I maintained about 120 KIAS because of a right crosswind condition. Over the threshold, I slowly reduced power to flight idle and initiated the flare.

Despite being trimmed, the aircraft exhibited the heavy flare forces I anticipated. But I was able to land with only my left hand on the control wheel. After touchdown, I lowered the nose gear to the runway, lifted the power levers up and pulled them aft into beta range to reverse propeller blade angle.

I taxied back to the runway threshold for departure. The 1900D climbed to 14,500 ft., and I established a cruise power setting of 93% N_g, 1,400 N_p, and 3,000 ft. lb. of torque for our flight back to Beech Field. Fuel flow was 420 lb. per hr. on each engine.

En route, I wanted to explore the 1900D's slow flight and stall characteristics in both the cruise and approach-to-landing configurations. After making clearing turns, I selected landing gear DOWN and LANDING flaps. The gear and flaps caused only minor pitch changes, and I quickly eliminated the higher elevator forces by using the electric trim system.

To maintain altitude, I added back pressure and retrimmed as airspeed decreased. Like its King Air brethren, the 1900D exhibits the benign, predictable slow-flight handling traits that are traditional Beechcraft hallmarks. As the aircraft approached stall speed, ailerons, rudder and elevator retained excellent control authority.

The aerodynamic buffet began at about 95 KIAS and the stall occurred at 80 KIAS. Although I held the control wheel aft to keep the aircraft stalled, the 1900D remained friendly and merely mushed downward. The aircraft resumed flying as soon as I released back pressure and reduced angle of attack,

I added power and with a positive climb rate established retracted the landing gear and flaps. About 2,000 ft. was lost in the maneuver. After climbing to 14,500 ft., I reduced power for a second stall with gear and flaps up.

As before, the aircraft responded well to control inputs as speed decreased. The buffet occurred at 100 KIAS and the wings stalled at 90 KIAS. The break was more gentle than the previous stall, but completely predictable. As with other Beech aircraft from the Bonanza to the Model 350, the pilot is given a wealth of aerodynamic signals—in addition to the warning tone—that the wing is about to stall.

At a gross weight of 16,100 lb. with 35 deg. landing flaps, the 1900D will stall at 82 KIAS, Hildreth said. The 1900D offers no surprises in the stall or when flying at low speeds, and gives the pilot more than adequate warning that the stall is imminent.

To hasten our return to Beech Field, I increased power to achieve a high-speed cruise of 286 KIAS while burning 520 lb. of fuel hourly per engine. Wichita Approach Control vectored us for a visual approach to Runway 36.

As the aircraft descended, I reduced power further to 180 KIAS before entering the traffic pattern. A wind from the east at 20 kt. would present a good opportunity to evaluate the 1900D's crosswind capability.

Landing gear and 17 deg. of flaps were selected on downwind as I retrimmed the aircraft. Hildreth verified a V_{ref} of 111 KIAS for the approach. Landing flaps were selected on final and I maintained about 120 KIAS to compensate for the crosswind.

With the runway assured, I slowly reduced power over the threshold, added right aileron to keep from drifting laterally, pushed left rudder to align the aircraft's nose with the runway centerline and initiated the flare.

As I expected, the 1900D responded predictably and quickly to my control inputs. The right main gear touched down, followed a few seconds later by the left gear and then the nose tire. With the aircraft on the runway, I applied full right aileron into the wind, lifted the power levers up and aft into beta range to aid deceleration and applied braking.

After the aircraft had slowed to a safe speed, I moved the power levers forward to flight idle but continued braking. I taxied off the runway and returned to the ramp for shutdown. During the flight 1,400 lb. of fuel was consumed.

In the tough, competitive regional airline market, Beech has a winner in the 1900D. Its more comfortable seating, stand-up headroom, EFIS and more powerful engines are major improvements over its predecessor. Those advantages, coupled with its relatively long range, high speed and proven systems reliability, make it a strong contender against other 19-seat aircraft.

From a pilot's standpoint, I liked the 1900D's docile, predictable handling and well-harmonized controls. From a passenger's standpoint, it has a large cabin that is reasonably quiet for a regional transport, and seats are designed for long flights of 2 hr. or more. From an operator's standpoint, system reliability and the PT6A engines contribute to decreased downtime and improved profits.

In the 10 years since the 1900 made its first flight, Beech has had to learn some tough lessons about the regional airline business in order to remain competitive. As a result, the company is striving to improve its product, service—and most importantly—customer support for airline aircraft.

With the advent of the 1900D regional transport, Beech has developed an advanced, 19-seat aircraft that can grow in the future and compete more effectively in the U.S. and the emerging global marketplace.

BEECHCRAFT 1900D SPECIFICATIONS

GENERAL

All-metal, twin-engine turboprop regional airline transport certified to Part 23 Commuter Category through Amendment 34 of the Federal Aviation Regulations. Approved for day/night VFR/IFR operations and flight into known icing conditions.

Minimum flight crew	Two
Base price (non-fleet purchase)	$4.37 million

POWERPLANTS

Two Pratt & Whitney Canada PT6A-67D turboprop engines, each flat rated to 1,279 shp. Two Hartzell full-feathering, fully reversible four-blade propellers with autofeather capability.

WEIGHTS

Maximum zero fuel weight	15,000 lb. (6,804 kg.)
Maximum ramp weight	17,060 lb. (7,738.4 kg.)
Maximum takeoff weight	16,950 lb. (7,688.5 kg.)
Maximum landing weight	16,100 lb. (7,302.9 kg.)
Useful load	6,510 lb. (2,952.9 kg.)

DIMENSIONS

Length	57.83 ft. (17.62 m.)
Height	15.48 ft. (4.71 m.)
Wingspan	57.83 ft. (17.62 m.)
Wing area	310 sq. ft. (28.7 sq. m.)
Wing aspect ratio	10.83
Wing loading	54.7/sq. ft. (267 kg./sq. m.)
Power loading	6.6 lb./shp. (32.2 kg./sq. m.)

CAPACITIES

Maximum usable fuel	665.4 gal. (4,484 lb./2,518.5 l.)
Baggage (aft)	175 cu. ft./1,630 lb. (4.9 cu. m./739.3 kg.)

PERFORMANCE

Maximum certified altitude	25,000 ft. (7,620 m.)
Maximum operating speed (V_{mo})	248 KIAS (459.5 km./hr.)
Rate of climb, two engines, sea level	2,625 ft./min. (800 m./min.)
Rate of climb, one engine, sea level	675 ft./min. (205.7 m./min.)
Range (25,000 ft.)	705 naut. mi. (1,306.3 km.)
Cruise speed (typical)	288 KTAS (533.7 km./hr.)
V_{mca}	92 KIAS (170.4 km./hr.)

Grand caravan features rugged airframe, simple systems befitting its utility role

Edward H. Phillips/Wichita, Kan.
December 2, 1991
Cessna Aircraft Co.'s spartan but rugged Grand Caravan combines a high degree of utility, mission flexibility and turboprop reliability with benign handling characteristics for a single-engine aircraft of its size and weight.

This *Aviation Week & Space Technology* pilot flew a company demonstrator aircraft during a 1-hr. 30-min. flight from Cessna's headquarters here. I was accompanied by Peter B. Hall, a Caravan demonstration pilot.

Cessna's claim that the Grand Caravan is the world's largest turboprop single-engine aircraft currently in production was reinforced during the preflight inspection. The aircraft has a wingspan of more than 52 ft., is nearly 42 ft. long and stands more than 14 ft. high at the rudder.

The fuselage features a 4-ft. plug 20 in. ahead of the wing and a 28-in. extension aft of the wing—modifications introduced on the 1986 Super Cargomaster, from which the Grand Caravan evolved.

Befitting the Grand Caravan's role as a cargo/utility aircraft, the fuselage has a two-piece cargo door on the left side and a smaller, passenger airstair-type door on the right. The aircraft will accommodate up to 14 occupants, including one pilot, or can be configured to haul cargo. Maximum useful load is 4,682 lb.

Cessna's Grand Caravan evolved from the Super Cargomaster, first introduced in 1986. Both aircraft feature a 4-ft. plug 20 in. ahead of the wing and another 28-in. stretch aft of the wing. The aircraft is certified for operation on floats.

Because Cessna has increased the Grand Caravan's maximum gross takeoff weight to 8,750 lb., the outboard flap segments are modified with specially designed Wheeler vortex generators and trailing edge Gurney angles to meet the Federal Aviation Administration's stall speed limitation for this class of aircraft. The devices promote attached airflow over the flaps at high angles of attack. By using them, Cessna was able to reduce stall speed at maximum flap deflection to the 61-kt. limit.

I inspected the Pratt & Whitney Canada PT6A-114A turboprop engine, which produces 675 shp. at 1,900 propeller rpm. Easy access to the engine and most accessories as well as the lead-acid battery are provided through the large, hinged cowl panels.

The fixed, tricycle landing gear is designed and built for rough-field operations. The main landing gear units absorb shocks through hollow tubular spring steel struts connected to a steel inter-tube, and have a heavier wall thickness than previous Caravan versions. The fixed nose gear uses an oil snubber and drag link spring.

The cockpit is reached through dual doors that swing out 180 deg. and secure in place. I entered the cockpit via a small stepladder that unfolds adjacent to the pilot seat. Hall did the same on the right side.

With my seat and safety belts adjusted, I found avionics, systems switches and controls within easy reach of the pilot—particularly the primary electrical system and engine start switches mounted on a vertical panel against the left sidewall. Overhead panels housing fuel tank select handles; additional switches and the standby flap motor control system also were easy to locate and reach.

Prior to engine start, the aircraft weighed 7,100 lb., including 2,254 lb. of fuel. Starting the PT6 engine was straightforward, and noise level was relatively low at an idle setting of 52% gas generator speed (N_g) and 1,000 propeller rpm (N_p). I taxied to Runway 19R at Wichita's Mid-Continent Airport.

Vision from the cockpit was excellent and the mechanical nosewheel steering was precise. I lifted the power lever up and moved it aft into beta range, decreasing propeller blade angle, to control taxi speed.

We completed the takeoff checklist, and extended the engine's inertial bypass vane to prevent foreign object ingestion during the takeoff roll. The vane deflects heavier-than-air objects away from the first-stage compressor blades, but causes ITT to rise about 20C and torque (T_q) to fall about 50 lb. The vane is used at pilot's discretion on the ground, and must be used in flight during icing conditions. Wing flaps were set to the recommended 20-deg. deflection.

The Grand Caravan has a two-piece cargo door on the left side and a smaller, passenger airstair-type door on the right. The aircraft seats 14, including the pilot. It may also be configured to carry cargo.

Grand Caravan's instrument panel and cockpit layout are spartan but functional. Central control stand houses engine, propeller and trim controls. Circuit breakers and engine start switches are installed in vertical panel on the left sidewall.

Cleared for takeoff, I held the brakes and advanced the power lever until more than 1,000 lb. T_q was attained and propeller speed (N_p) reached a maximum of 1,900 rpm. Upon brake release, the aircraft accelerated smoothly and required little right rudder to track the runway centerline.

I eased the control wheel aft at the 65-kt. rotation speed (V_r) and the Grand Caravan flew off the runway. Rotation forces were not heavy, and the aircraft accelerated quickly to the recommended climb airspeed of 120 KIAS. I trimmed the elevator to maintain that speed. Rate of climb was 1,400 ft./min.

With the wing located far aft of the cockpit, inflight visibility was excellent in all directions, and the long engine cowling did not block my view forward. During the climb to 8,500 ft., I made shallow banks for collision avoidance and found roll and pitch forces much lighter than I had anticipated.

The aircraft responded quite swiftly to control inputs, and generally handled like a large Cessna 182. At 8,500 ft. and 15C, a power setting of 1,350 lb. T_q, 335 lb./hr. fuel flow and 1,900 N_p provided a cruise speed of 150 KTAS.

In preparation for a stall series, I performed clearing turns at bank angles ranging from 30 to 60 deg. Both pitch and roll forces remained surprisingly moderate, and excessive elevator forces were easily trimmed away. The Grand Caravan's lateral and longitudinal stability was exceptionally good, both in steep turns as well as at shallow bank angles.

I decreased airspeed to below 125 kt., reduced power to flight idle and extended the single-slotted, semi-Fowler wing flaps to their maximum 30-deg. deflection. An aerodynamic buffet preceded the approach-to-landing stall, with the break coming at 50 KIAS.

Stalls with flaps up provided a less defined break, and were more benign than those with flaps down. After the stall break at about 78 KIAS, I was able to hold full aft elevator and keep the wings level by using rudder alone as the aircraft descended.

Recovery from the stall, whether in level flight or turns, could be accomplished simply by reducing angle of attack. Generally, the Grand Caravan's stalling characteristics closely resembled those of the piston-powered single-engine Cessnas.

With the stall series completed, Hall gave me a heading to fly toward the municipal airport at Kingman, Kan., for takeoffs and landings. I entered a left downwind for Runway 18 and slowed the aircraft to 100 kt. as Hall completed the landing checklist.

Easy handling in pattern

Flying the Grand Caravan in the traffic pattern was easy. I felt comfortable with the aircraft and appreciated its low pilot workload and excellent visibility from the cockpit. On final, I selected 30-deg. full flaps and trimmed for an 80-kt. approach speed. There was a 10-kt. crosswind from the southwest.

Over the runway threshold, I reduced power to flight idle, kept the right wing down with aileron, initiated the flare and added left rudder to align the aircraft's longitudinal axis with the runway centerline. Control forces were not heavy, and the Grand Caravan responded quickly to my inputs.

Although the approach had gone well, I misjudged the flare and the aircraft hit too hard on the right main gear and bounced back into the air about 5 ft. before settling on the runway to stay. I added full right aileron into the wind, lifted the power lever up and moved it aft to select the maximum reverse propeller blade angle of –18 deg. The aircraft took my bumbled landing in stride, once again exhibiting the same forgiving nature of its smaller Cessna siblings.

I made an additional five landings at Kingman to become better acquainted with the aircraft before flying to Maize airport southeast of Wichita for short-field takeoffs and landings. Maize's single runway is often used by Cessna pilots to demonstrate the Grand Caravan's short-field capabilities and rugged landing gear. Runway 17-35 is a 2,100-ft. turf strip bounded on each end by roads.

I entered a left downwind for Runway 17 and selected full flaps on final. Although Maize is an easy airport for the Grand Caravan to master, it requires extra care and planning by the pilot to use safely. For example, Hall reminded me to watch for tractor-trailer trucks coming down the road. If, during the approach, it appeared that a truck would be passing by the runway threshold at the same time we would be flying over it, we would execute an immediate go-around. I agreed completely.

Fortunately, there were no trucks on my first approach, only a slight crosswind from the southwest. Over the road, I reduced power to flight idle and initiated the flare. The aircraft touched down firmly without bouncing. I selected full reverse, retracted the flaps and applied hard braking. We used about two-thirds of the runway to decelerate before taxiing back for takeoff.

After lining up with the runway, I selected the recommended 20 deg. of flaps for departure, held the brakes and smoothly added power. As torque passed through 1,000 lb., I released brakes and continued adding power up to about 1,600 lb. T_q.

At 70 kt. I rotated the aircraft and maintained the recommended 83-kt. climb speed to clear the road and trees at the opposite end of the runway. After the next landing, Hall suggested using only 10 deg. of flaps for takeoff to experience the difference in pitch forces during rotation and climbout.

The lack of 10-deg. additional flap deflection significantly altered the aircraft's takeoff handling. At 70 kt., I moved the control wheel aft to rotate the aircraft, but the nose remained on the ground for about 2 or 3 sec. before sluggishly moving upward.

Main gear liftoff occurred about two seconds after rotation, with less than 500 ft. of runway remaining. I attained and maintained 83 kt. as we passed over the road. I made two more takeoffs and landings using normal flap settings before flying back to Mid-Continent and Cessna's facilities.

The company began delivering Grand Caravans in September, 1990, and has delivered about 14 aircraft to date. Cessna has a production backlog into 1993, according to William C. Hogan, manager, propjet marketing support.

I enjoyed flying the Grand Caravan. It demands no special piloting skills for an aircraft of its size, and its simple systems, rugged airframe and reliable PT6 engine are key factors in its success as a utility aircraft.

CESSNA 208B GRAND CARAVAN SPECIFICATIONS

GENERAL

- All-metal, 14-seat turboprop monoplane utility and business aircraft.
- Normal category certification, Part 23 of the Federal Aviation Regulations.
- Approved for day/night VFR/IFR flight operations, including flight into known icing conditions.

Minimum flight crew	One
Base price (typically equipped)	$1.025 million

POWERPLANT

One Pratt & Whitney Canada PT6A-114A reverse flow, free-turbine turboprop engine, 675 shp. One Hartzell constant-speed, full-feathering, reversing propeller, 100 in. diameter.

WEIGHTS

Empty weight	4,103 lb./1,861.1 kg.
Maximum ramp weight	8,785 lb./3,984.8 kg.
Maximum gross takeoff weight	8,750 lb./3,968.9 kg.
Maximum landing weight	8,500 lb./3,855.5 kg.
Maximum useful load	4,682 lb./2,123.7 kg.

DIMENSIONS

Length	41.7 ft./12.7 m.
Height	14.1 ft./4.3 m.
Wingspan	52.1 ft./15.9 m.
Wing area	279.4 sq. ft./25.9 sq. m.
Aspect ratio	9.5

Wing loading	31.3 lb./sq. ft.
Power loading	13 lb./sq. ft.

CAPACITIES

Usable fuel capacity	332 gal./1,256.6 l.
Cabin volume (aft of cockpit)	340 cu. ft./9.63 cu. m.

PERFORMANCE

Maximum approved operating altitude	25,000 ft./7,620 m.
Maximum approved operating altitude (icing conditions)	20,000 ft./6,096 m.
Cruise speed (10,000 ft.)	182 kt./337 km./hr.
Rate of climb, sea level, gross weight	975 ft./min./297.2 m./min.
Range (10,000 ft.)	900 naut. mi./1,666.8 km.

Engine boosts Jetstream performance

David M. North/Washington
January 3, 1983

The installation of Garrett TPE331-10 turboprop engines on the newest version of the British Aerospace Jetstream gives the corporate and commuter aircraft increased reliability and performance that complement the aircraft's straightforward handling characteristics and large cabin.

First production version of Jetstream 31, with British registration G-TALL, was flown here recently by this *Aviation Week & Space Technology* pilot. The aircraft's two TPE331-10 turboprop engines are rated at 900 shp. each at takeoff.

Prior to the flight from Washington's Dulles International Airport, Douglas McDonald, flight operations manager for British Aerospace, detailed some of the changes that had been incorporated in the Jetstream 31 from earlier versions. The biggest change has been the installation of the Garrett engines. The Jetstream 31 also has Dowty-Rotol four-bladed propellers.

The electrical system of the aircraft was changed from alternating current to a 28-v.d.c. system with an alternating current subsystem. The electrical change was made to match the output of the Garrett engine components.

British Aerospace's G-TALL Jetstream 31 is the first production aircraft from the company's Prestwick, Scotland, manufacturing facility. Aside from the installation of the Garrett TPE331-10 turboprop engines, with the air intake located above the engine, the exterior of the commuter and corporate aircraft closely resembles the Jetstream that was first developed by Handley Page in the mid-1960s.

The hydraulic pump has been changed, but it is the same simple and redundant system incorporated in the earlier aircraft. The aircraft's double-slotted flaps are hydraulically operated. The primary flight surfaces are connected by mechanical linkage with the controls in the cockpit.

The exterior of the Jetstream 31 has changed little from its predecessors, but the cockpit instrumentation has been changed to incorporate newer technology. Collins Pro-line avionics are standard on the aircraft. The forward instrument panel is designed to minimize pilot workload and is not cluttered. Many of the system controls are located in the aircraft's overhead panel.

The Jetstream 31's wing was modified to accommodate the increase in maximum takeoff gross weight from 12,500 lb. to the 14,500 lb. allowed under current U.S. regulations for an aircraft of the Jetstream's category. The Jetstream 31 was certificated to the Federal Aviation Administration's Special Federal Aviation Regulations 41 on Nov. 30. The aircraft had been issued a certificate of airworthiness by the British Civil Aviation Authority earlier this year.

As I sat in the left seat of the Jetstream 31, McDonald discussed some of the avionics options while in the right seat. The commuter version of the aircraft can be equipped with a Sperry 200B automatic flight control system as optional equipment. British Aerospace is certificating and recommending the corporate version of the aircraft equipped with the SPZ 500 flight control system. The Collins WXR300 color radar is standard for the commuter aircraft, and a checklist can be displayed by the radar system, if desired.

The aircraft we flew is equipped with a yaw damper, but McDonald said the system is not required. The demonstrator flown from Dulles also was equipped with an 80-lb.-force stick pusher in the event of a stall, but this system is to be removed in later production aircraft. The removal of the stick pusher is allowed because of the aircraft's straightforward stall characteristics, McDonald said.

Once McDonald finished the prestart checklist, the two Garrett engines were started with the aid of an external start cart. British Aerospace and Garrett are evaluating the possibility of installing an auxiliary power unit in the Jetstream, especially for corporate use.

The ramp weight of the aircraft was calculated at 12,000 lb., including 2,600 lb. of fuel, two pilots and one passenger. The maximum fuel capacity of the aircraft is 3,079 lb. The ramp weight of G-TALL was 88.7% of its maximum 14,660-lb. ramp weight.

The V_1 takeoff decision speed was calculated at 105 kt. by McDonald. The V_r rotation speed and the V_2 takeoff safety speed was

108 kt., according to the aircraft's flight handbook for the gross weight and the 15C conditions at the time of the flight.

During taxi from the Page Airways ramp to the relatively close departure end of Runway 19L at Dulles, the Jetstream's nosewheel steering wheel at the pilot's left side was easy to operate and responsive even at low speeds. At no time during the taxi to or the return from the runway was there any evidence of nosewheel shimmy. There also was little if any oscillation of the nose during application of power or brakes.

Final takeoff power was set by McDonald after I had moved the power levers to an approximate setting. The nosewheel steering was used to approximately 50 kt., when the aircraft's rudder became effective. Takeoff roll was approximately 2,000 ft., prior to liftoff.

Within 5 min. the aircraft had a rate of climb of 2,000 fpm passing through 5,000 ft., with a fuel flow of 395 lb./hr./engine. The climb speed was 150 kt. indicated. British Aerospace recommends a long-range climb speed of 140 kt. calibrated and a maximum climb speed of 160 kt.

The 2,000-fpm rate of climb was maintained through 10,000 ft. At that altitude, the fuel flow was 370-lb./hr./engine at a speed of 152 kt. At 14,000 ft. the speed of the Jetstream had increased to 170 kt. and the rate of climb declined to 900 fpm.

The maximum 650C internal turbine temperature was being maintained by the Garrett engine's temperature and torque limiter computer during the climb. The final cruising altitude of 16,000 ft. was reached 20 min. after takeoff. However, lower altitudes assigned by the Washington air traffic control center for short periods of time made it impossible to compare the aircraft's climb performance with figures in the flight handbook.

McDonald had chosen 16,000 ft. as a representative altitude that would be flown by commuter operators on relatively short route segments. The Jetstream 31's maximum certificated altitude is currently 25,000 ft., but British Aerospace is considering having the altitude raised to 31,000 ft., especially for corporate operations.

Flight parameters

The fuel used for the taxi, takeoff and climb to 16,000 ft. was approximately 200 lb. Once the aircraft was level at 16,000 ft., the power was left at 650C and close to an 80% torque setting. The indicated speed of the Jetstream 31 was 205 kt., and on the warmer-than-standard conditions at that altitude, the true airspeed of the aircraft was calculated to be approximately 295 kt. Fuel flow at the maximum cruise power setting was 330 lb./hr./engine.

Pulling the power back to a 70% torque setting with 97% rpm resulted in a 190-kt. cruise speed and a 300 lb./hr./engine fuel flow.

While still at 16,000 ft., 45- and 60-deg. bank turns were made to obtain a feel for the responsiveness of the aircraft at that altitude. The roll rate of the aircraft was found to be quick during the turns. The Jetstream has 15-deg. of rudder movement coordinated with the aileron to facilitate turns. Pitch control also was responsive during the entire flight.

Stalls accomplished

Two stalls were accomplished at 15,000 ft. while the aircraft was in the clean configuration. The stick shaker activated at 100 kt., and the stall and stick pusher activation occurred simultaneously at 90 kt.

In both maneuvers the stalls were straightforward, with no tendency to lose aileron or elevator control approaching the stall. In the first stall, the aircraft favored a left wing down attitude, while in the second the right wing dropped slightly before being corrected with aileron.

The noise level was low in the cockpit during the cruise at 16,000 ft. and while in the landing pattern, and McDonald and I were able to discuss the performance of the aircraft at conversational levels. The visibility from the cockpit also was excellent during the flight and the two landings made on Runway 19R at Dulles.

The descent to the landing pattern was made at 170 kt. at a fuel flow of 150 lb./hr./engine. The Jetstream maintained a consistent rate of descent with wings level and without the use of the autopilot or hands on the control column.

Another demonstration of the stability of the aircraft occurred as we passed through some minor turbulence. The turbulence was absorbed by the wings with little of the motion transmitted to the fuselage. Little control input was needed to hold the aircraft on its desired heading and rate of descent.

Fuel flow in the landing approach at 150 kt. was 280 lb./hr./engine. The landing reference speed for the Jetstream 31 was calculated to be 105 kt. at the 11,400-lb. landing weight.

Both the touch-and-go and the final landing were made in no-wind conditions. The first approach was made using the instrument landing system at Dulles. The Jetstream 31 was responsive to power and control changes during corrections to altitude and speed.

The second approach made to a final landing was under visual flight conditions at 1,000 ft. to stay below the cloud cover. Both landings were smooth, as I had little difficulty adjusting to the eye reference height from the cockpit of the main landing gear. A total of 625 lb. of fuel was used during the 1.4 hr. flight from engine start to shutdown.

The Jetstream 31's airframe and systems have been designed with simplicity and reliability taken into consideration. However, British Aerospace has chosen not to incorporate the newest technology available in construction, systems or avionics into the new version aircraft. The performance and handling characteristics of the commuter and corporate aircraft reflect this philosophy, with no surprising adverse traits inherent in the aircraft.

British Aerospace intends to produce 18 Jetstream 31s in 1983, followed by 25 in 1984. The company has targeted the manufacture of 36 aircraft for 1985 and beyond, but that annual rate may be corrected, either up or down, depending on market conditions.

Engine changes mark Jetstream evolution

Washington

Evolution of the Jetstream series from the first aircraft developed by Handley Page in the mid-1960s to the 18- or 19-passenger aircraft delivered by British Aerospace late last year has been marked by numerous changes of engine manufacturers.

The Jetstream's basic airframe has remained the same during the past two decades, while at least three different engine manufacturers have been associated with the program, either in initial production or with retrofit programs. The initial Jetstream was powered by two Societe Turbomeca Astazou turboprop engines. Other versions of the Jetstream were powered by earlier Garrett TPE331 turboprop engines. Pratt & Whitney of Canada PT6A turboprop engines have been retrofitted on some of the Jetstreams built in the late 1960s and early 1970s.

British Aerospace's G-TALL demonstrator has a 12-passenger executive shuttle configuration. Other interiors offered as standard by British Aerospace are an eight- or nine-passenger executive configuration and an 18- or 19-passenger commuter version.

The price of the 18-passenger commuter Jetstream is approximately $2.25 million in 1981 dollars. With escalation and the installation of additional avionics packages for the commuter aircraft, the price is expected to be close to $2.5 million for delivery in 1983.

The commuter version of the Jetstream can be configured with either an 18- or 19-passenger interior. The 18-passenger aircraft has a 32-in. pitch and the option of a toilet. The 19-passenger version has a 30-in. seat pitch on one side and the toilet is not included. Both versions have an offset aisle to accommodate the two-and-one seating. The commuter interiors will be done by Field Aircraft Services.

The executive-configured Jetstream can be completed at any location selected by the operator. The price of the executive-configured aircraft is expected to be approximately $3 million, after increased avionics and a corporate interior are added.

British Aerospace's marketing team expects sales of the Jetstream will be divided nearly equally between commuter and corporate operators. The lack of any sales so far of the Jetstream 31 to U.S. commuter or corporate operators is viewed as a reflection of the recessionary economy rather than any shortcoming of the aircraft.

At least one U.S. commuter operator had signed for the Jetstream 31 but has since declined to accept the aircraft because of a decline in traffic. British Aerospace believes the reluctance of corporate operators to place an order for the Jetstream 31 was primarily because a demonstrator has not been available until this year. The recent tour of the U.S. by two aircraft is expected to generate sales in the corporate market.

Company officials believe the biggest selling point of the Jetstream 31 is the size of its cabin, especially the 5.9-ft. headroom. The headroom is greater than that of its two main U.S. competitors—the Beech Aircraft 1900 Airliner and the Fairchild Aircraft Metro 3. Both these aircraft also are pressurized. "In the Jetstream 31's cabin you have the same feeling of space as in the jet transports," David Field, British Aerospace's director of marketing-airlines, said.

The choice of the Garrett TPE331-10 turboprop engines over Pratt & Whitney of Canada PT6A engines for the Jetstream 31 was discussed by Kenneth Spinney, British Aerospace's vice president of marketing.

"In our talks with the commuters, we find a reluctance by 50% of them to buy an aircraft with Garrett turboprop engines installed," Spinney said. "However, we chose the TPE33 1-10 because of its performance and reliability, and it will turn out to be an excellent engine. Much of the commuter opposition to the engine is based on the TPE331-3 engine and its earlier problems, which have since been corrected."

Another factor in the choice of the Garrett engine by British Aerospace is that its predecessor, Handley Page, had fitted earlier Garrett TPE331 turboprop engines on a prototype aircraft for a U.S. Air Force aircraft program that later was terminated.

JETSTREAM 31 SPECIFICATIONS

POWERPLANT
Two Garrett TPE331-10UF-501H turboprop engines rate at 900 shp. each at takeoff.

PROPELLERS
Two Dowty-Rotol four-bladed, 106-in.-dia. with full feather and full reversible capabilities.

WEIGHTS	
Maximum ramp weight	14,660 lb. (6,650 kg.)
Maximum takeoff weight	14,550 lb. (6,600 kg.)
Maximum landing weight	14,550 lb. (6,600 kg.)
Maximum zero fuel weight	13,228 lb. (6,000 kg.)
TYPICAL OPERATING WEIGHT	
Commuter	9,046 lb. (4,103 kg.)
Executive shuttle	9,482 lb. (4,301 kg.)
Corporate	9,613 lb. (4,360 kg.)
Fuel capacity	3,079 lb. (1,397 kg.)
DIMENSIONS	
Length	47.2 ft. (14.4 meters)
Height	17.5 ft. (5.3 meters)
Wingspan	52 ft. (15.8 meters)
Cabin headroom	5.9 ft. (1.8 meters)
Cabin length	24.3 ft. (7.4 meters)
Cabin width	6.1 ft. (1.9 meters)
Cabin volume	598 cu. ft.
DESIGN CONDITIONS	
Maximum operating altitude	25,000 ft. (7,620 meters)
Maximum operating speed	230 kt. indicated
Maximum operating Mach	0.475
Cabin pressurization	5.5 psi.
PERFORMANCE	
Range in instrument conditions, standard day	
Commuter configuration with 19 passengers	630 naut. mi.
Executive shuttle with 12 passengers	950 naut. mi.
Corporate version with nine passengers	1,150 naut. mi.
Takeoff distance, accelerate/stop	4,150 ft. (1,265 meters)
Landing field length	2,215 ft. (675 meters)
Twin-engine climb rate	2,200 ft./min.
Single-engine climb rate	510 ft./min.

Note: Ranges with visual flight reserves are approximately 200 naut. mi. farther.

Index

Illustrations are in **boldface**